Supporting Play-Based Learning in Primary Mathematics Curriculum

Mehmet Raci Demir

Supporting Play-Based Learning in Primary Mathematics Curriculum

Effect of Teaching Mathematics through Different Play Types on 1st Grade Students Achievements and Retention Levels

**Bibliographic Information published by the
Deutsche Nationalbibliothek**
The Deutsche Nationalbibliothek lists this publication in the Deutsche Nationalbibliografie; detailed bibliographic data is available online at http://dnb.d-nb.de.

Library of Congress Cataloging-in-Publication Data
A CIP catalog record for this book has been applied for at the Library of Congress.

ISBN 978-3-631-82976-9 (Print)
E-ISBN 978-3-631-84604-9 (E-PDF)
E-ISBN 978-3-631-84605-6 (EPUB)
E-ISBN 978-3-631-84606-3 (MOBI)
DOI 10.3726/b17983

© Peter Lang GmbH
Internationaler Verlag der Wissenschaften
Berlin 2020
All rights reserved.

Peter Lang – Berlin · Bern · Bruxelles · New York · Oxford · Warszawa · Wien

All parts of this publication are protected by copyright. Any utilisation outside the strict limits of the copyright law, without the permission of the publisher, is forbidden and liable to prosecution. This applies in particular to reproductions, translations, microfilming, and storage and processing in electronic retrieval systems.

This publication has been peer reviewed.

www.peterlang.com

To my sons Salih Eren and Onur Akif. ‥ .

Foreword

One of the main problems of Turkish education system is that students are generally anxious about math class, dislike or cannot succeed it. When school life is taken into consideration, children's problems with math class, their negative attitudes, anxieties and failures are immense in national and international examinations. Among the reasons of this situation, it is observed that there is a change in the reactions of children that can solve mathematical operations from the earlier ages to math class in their school life.

According to the research carried out on these issues, many students around the world show fear and antipathetic response to math class, and many students have difficulties in understanding and learning math. Anxiety for math class increases throughout the time, also interest in math decreases in the progress of time. Students have no problems with math in the initial years of school or have little issues, but throughout the years passing by, there are declines in their achievement in math.

The reason why students feel nervous in mathematic lessons is because teachers cannot convey or teach students the mathematical concepts. The cause and effect relation in math or the impact of math in students' real lives sufficiently or enough. Teachers are supposed to explain the purpose of math and its actual relation with real life and bring students in the logic of mathematical operations by including activities to attract students in their classes. On the other hand, in math classes taught via plays, students state that they do not feel they are in math class, and their fear fades away. They also indicate that they enjoy participating in classes held via teaching through plays or as if plays, and they do not forget what they have learnt for a length of time.

To eliminate this problem, in this study, the impact of teaching through different types of plays in math class of first graders at primary school has been examined. Students' points of view about learning math through plays have been studied. If teaching through plays becomes effective, teachers and students will be taught how to involve plays, the principles of the method and its strengths and weaknesses in math class.

In this way, it is thought that beginning from the first grade, negative attitudes, fear and failures would be blocked. Furthermore, it is thought that using the method of teaching through plays in math classes would set an example for further research that will be carried out by presenting theoretical and practical sample plays and activities.

Mehmet Raci Demir,
Istanbul,
August, 2020

Preface

From the first years of my teaching experiences, I spent my knowledge in theory, thinking about how to put it into practice and trying different methods. My students mostly had a low socio-economic level, and their misbehavior was higher than in schools in the other regions of city. They were talking about the boringness, difficulty and uselessness of the lessons. The learning centers I studied in my master's thesis offered them different activity-based environments, providing them with opportunities to be questioning and problem-solving individuals. Still, I felt there were a lack and an inadequacy because activities had limited impact. It was very difficult to keep their interest and attention, not to lose their focus, to make them learn the subject and to ensure retention of learning. I was trying to achieve these problems through shifting between centers. After I started my doctorate thesis, I realized that I did not think like children, which was the most important deficiency in my teaching. The children like play-oriented and play-centered activities and they live by play. As I examined the studies in the literature, I realized that research on learning and research on play are in parallel with each other. With changing paradigms, play and learning approaches have also changed. Examining mathematics teaching through plays has been another challenge. Moreover, how different could it be from primary school curriculum to learn mathematics and do operations by playing, with first grade students who have just learnt to read?

In this book, and in my doctoral dissertation study, the effect of teaching mathematics through different plays on the achievement and retention of first grade students was examined. In addition, the research was supported by the opinions of the students who played the games. Intra-class and outdoor games/plays that will contribute to researchers and teachers were supported by student evaluations, researcher observations and comments of the plays. Plays in this book were enriched with play research carried out in the world and in Turkey. Further studies will hopefully be more play based.

Acknowledgment

Firstly, I would like to express my sincere gratitude to my advisor Prof. Dr. Füsun Alacapınar for the continuous support of my Ph.D study and related research, for her patience, motivation and immense knowledge. I would like to express my sincere gratitude to Prof. Dr. Veysel Sönmez who made me love teaching and science, encouraged me and showed me new horizons for future. Their guidance helped me in all the time of research and writing of this book.

Besides my advisor, I would like to thank the rest of my thesis committee for their insightful comments and encouragements.

To make this study more qualified, my sincere thanks also go to my friends Dr. Selda Özer, also Asst. Assoc. Dr. Hülya Yildizli who does not refuse to support all my studies, and to my dear colleague Seda Yeşildal, who translated the book in English and to the editorial board of Peter Lang, Esra Bahsi.

I would like to thank my family: my parents and to my brothers and sister for supporting me spiritually throughout writing this book and my life in general.

Finally, I would like to thank my dear wife, whose support I always feel by my side with patience and understanding, and also to my sons Onur Akif and Salih Eren, who guided me with their plays.

Table of Contents

Introduction 17

Chapter 1 Play and Pedagogy of Play 23
  Play 23
  Historical Development of Play 23
  What Is a Play? 25
  Criteria of Plays 26
  Play Approaches 28
    A. Classical Play Theories 29
      1. Surplus Energy Theory 29
      2. Relaxation Theory 29
      3. Recapitulation Theory 29
      4. Pre-Exercise Theory 30
    B. Modern Play Theories 30
      1. Psychoanalytic Play Theory 30
      2. Arousal (stimulation)/Change of Pleasure Theory 30
      3. Meta-Communicative Play Theory 30
      4. Cognitive Play Theory 30
  Types of Plays 32
    Cognitive Plays 33
    Creative Plays 33
    Social Plays 33
    Manipulative Plays 34
  Principles of Developing Pedagogy of Play 35
    1. Beginning a Play 35
    2. Expanding a Play 36

3. Arranging Physical and Social Environment in Plays 37
4. Exploration, Play, Learning and Teaching Programs 39
Characteristics of Teaching through Plays 44

Chapter 2 Play and Mathematics Instruction 47

Teaching Mathematics through Plays 47
1. Mathematics Embedded in Play 48
2. Play Centering on Mathematics 49
3. Play with the Mathematics That Has Been Taught in School 50

Play Types in Teaching Mathematics 55
1. Manipulative Games 56
2. Card Games 58
3. Mind Games 58
4. Computer Games 59
5. Musical and Rhythmic Plays and Sports 59

The Advantages and Disadvantages of Teaching Mathematics
through Play 60

Teacher's Responsibilities in Teaching through Plays 62
1. Planning and Creating the Learning Environment That Requires
   Challenge 62
2. Enriching and Improving Children's Vocabulary through
   Communication in Plays 63
3. Supporting Children's Learning by Free and Structured Plays 64
4. Observing and Evaluating Children's Plays 65

Research on Teaching Math through Plays in Turkey 68
Research Abroad on Teaching through Plays 70

Chapter 3 Method 79

Research Design 79
Study Group 79

Data Collection Tools 81
Data Analysis 82
Intervention in the Research 83
    1. Truck-Loading Game 84
    2. Number Cubes (Find the Sum) 85
    3. Number Cubes (The Sum Is the Same) 89
    4. Card Games (Find the Number Pairs) 90
    5. Card Game (Find the Sum) 92
    6. Numbers at the School Garden (Mental Calculation; Addition) 93
    7. Numbers at the Garden (Mental Calculation, Subtraction) 95
    8. Darts (Mental Calculation Addition and Subtracting) 96
Findings 97
    1. Findings Related to First Hypotheses 97
    2. Findings Related to Second Hypotheses 98
    3. Findings Related to Third Hypotheses 98
    4. Findings Related to Fourth Hypotheses 99
Findings Related to Sub-Problems 100

Chapter 4 Conclusion, Discussion and Suggestions 109
4.1. Conclusions Related to the First Hypothesis and Discussions 109
4.2. Conclusions Related to the Second Hypothesis and Discussions 110
4.3. The Conclusions Related to the Third Hypothesis and
     Discussions 111
4.4. Conclusions Related to the Fourth Hypothesis and Discussions 111
4.5. Conclusions Related to Sub-Problem and Discussions 112
Suggestions 114

Appendix 1: 1. Anticipated Achievements in the Current Curriculum of Math Class for the First Grade 117

Appendix 2: First Grade Math Class Achievements in Experimental Group Implementation 119

List of Figures 123

List of Tables 124

References 125

Introduction

Culture can be defined as the life style of a society and a dynamic whole of feelings, thoughts and deeds of human beings apart from those carried out by the nature (Sönmez, 2007:5). Bringing individuals in cultural values is called enculturation. Education can be identified as enculturation in general sense, namely, the process of bringing individuals in cultural values (Demirel, 2004, 7–8).

Education may have different descriptions based on each philosophy. According to Ertürk (1972), education is described as "the process of causing intentional, terminal behavioral change in an individual's behaviors via his/her own experience", while Sönmez expressed it as "the process of creating desired, terminal bio-chemical changes in brain as a result of physical impulses" (Ertürk, 1998; Sönmez, 2007:5). It may be claimed that target bio-chemical changes can be realized in systematic time periods planned and programmed such as inspection, assessment and enhancement that sense organs are set on. Those processes can be gathered up under the title "the process of education and training."

The process of education and training can be approached as an open system to realize the change mentioned above. The open system includes input, output, feedback and operations digit. Operation digit is the part where inputs are formed in the direction of objectives by the means of proper and effective chemical, physical, mental and operational processes. Shortly, the position of education is also described as the arrangement of setting and environment (Sönmez, 2007; Fidan, 1996).

The position of education or educational status can be defined as "teaching activities prepared, applied, evaluated and improved in order to bringing one in terminal behaviors." Through educational status, "it is possible to transform a student who does not know into a student who knows, a student who does not do into a student who does, a student who does not like into a student who likes, a student who does not have democratic attributes and behaviors into a student who has them" (Sönmez, 2007:87).

Educational status may have a dynamic constitution originated as a result of physical features of teaching environment, equipment, educational technology, time, reinforce, feedback, correction and hint, love, qualification of teacher, student's involvement, student's readiness level and their interaction with proper teaching and learning methods. Several issues are encountered in the interaction of these factors. One of these issues is the matter how the behaviors in the direction of objectives can be gained by students via educational status. This situation makes the problem of teaching methods a current issue (Gömleksiz, 1993: 15).

There have been many definitions for the concept of method. Generally, method can be defined as "the shortest route to the objective" (Fidan, 1996:168) or "a straight path chosen to teach a subject" (Demirel, 2004:72). In addition, it is stated that technique is included in the concept of method (Demirel, 2004 and Gözütok, 2007). Teaching method can also be described as the method followed to present teaching materials and sources, construct teaching activities, a way of implementation of a particular teaching method or a sort of study meant for teaching activities management (Gömleksiz 1993:20–21). In the light of various explanations, it is stated that there is no single magical method for a teacher to make the students gain a subject, skill or behavior (Küçükahmet, 1995:37; Gömleksiz, 1993:18 and Gözütok, 2007:205). This situation may result from various reasons:

1. It is not likely for every student to learn via the same method.
2. A single method does not solely fit for all subjects.
3. A teaching method is not sufficient to realize all of the objectives determined.
4. All the teachers are not able to use all methods skillfully. Some teachers are prone to certain methods.
5. Some methods require a long time.
6. Some methods require monetary resources.
7. Some methods require certain physical circumstances.
8. Students are not attracted by all methods equally (Gözütok, 2007:205).

If teaching comprehension is appropriate with the needs and objectives of students, attendance to lessons and learning level of students increase. In this case, the content and comprehension of teaching should be arranged to meet students' expectations and make them achieve their purposes. One of the reasons to prefer different teaching methods may be that some methods also fulfill the need of amusement for students. In this context, it can be provided for students to learn entertainingly through experience by the means of educational plays. Moreover, plays can be useful for concentration, high level of class attendance and providing reinforcement (Uğurel, 2003:27). The most productive learning situations for students are the ones when students are active. As long as students are involved in activities in the process of learning and teaching, permanent learning is ensured. For Aytekin (2001), plays derive attention more than any other teaching methods since plays transform students into active state from passive one considering the students' attendance to the class. In this sense, when teachings via educational plays are examined, it can be said that studies are done on various age groups and in different disciplines.

The concept of play may have different definitions regarding each philosophy. Lazarus (1883) asserts that 'play is a free time activity that has no objective, emerges spontaneously and brings joy.' According to Hall (1906), a kid reflects the cultural improvement of humanity when he/she plays. Groos (1899) considers plays as preliminary tests for the maturity reached at the end of childhood (Songur, 2006:32). For Bilen (1999), play is defined as 'activities that improve physical and mental skills of an individual, make life amusing, and develop artistic, aesthetical qualities.' In this sense, putting aside the fact that plays maintain their importance in every step of life, they are considered to be indispensable, essential learning environment in childhood when physical, social, cognitive, psychomotor and psychological developments are specifically important. During plays, children organize their own living spaces using their imagination. By raising questions and trial and error, what is learnt is reinforced in mind and gains a schematic structure. By the means of the plays, children improve abstract thinking skills. Children who conclude certain implications through the settings and narration in plays reflect many of those implications in their real lives (Uğurel, 2003).

Together with preschool education, plays continue more systematically and programmed in a child's life. Intra-class and outdoor plays played with friends take the place of plays played in family and on streets. It can be said that children do not benefit enough from enjoyable and instructive opportunities of plays considering the restricted play times and small classrooms arranged with desks lined one after the other at school life where they face various responsibilities for the first time. Beginning from the first years of primary school, plays can find place in break times through children's own effort, in non-class activities and P.E. classes. In secondary school and following years, it can be said that plays are considered to be out-of-school and non-class activities or meant for enjoying oneself rather than being instructive. No matter what the purpose of constructing a play is, education through plays can be applied to all age groups.

Plays can be said to be used as the methods and techniques to realize terminal objectives when applied to different disciplines. The fact that structures, rules and purposes of plays change provides plays to be used as a method in teaching (Uğurel, 2003). When the literature is reviewed, it is seen that there have been researches on teaching through plays at primary and secondary schools in different disciplines. In this context, about teaching through educational plays at primary schools, it is seen that researches have been carried out via different variances in life sciences class (Bayazıtoğlu, 1996), in social studies class (Karabacak, 1996; Pehlivan, 1997), in maths (Altunay, 2004; Kılıç, 2007; Tural, 2005; Gelen and Özer, 2010), in science and technology class (Ercanlı, 1997; Ören and Avcı, 2004), first reading and writing class (Özenç, 2007), in visual arts class (Turanlı, 2012), in physical education class (Altun, 2013) and in music class (Değer, 2012). Besides, at secondary school education, researches in Turkish class (Gülsoy, 2013) and math class (Uğurel, 2003; Songur, 2006) can be examples for the usage of plays in different disciplines and age groups. In this context, reasons to prefer teaching through educational plays are stated by Samur (1983) as below:

1. Plays increases children's attention to the class.
2. Plays are a good motivation means for teachers.
3. During a play, a child gets to know his/her body and people and things around.
4. Children feel happy since they are active in plays.
5. While playing with different means, children get in contact with things around themselves.
6. Children take various roles during a play, they apprehend their social environment and state of people, and they impersonate them closely.
7. All games have certain rules to be obeyed. Children playing a game understand the rules easily and obey them.
8. While playing, children learn how to treat people.
9. Play enriches classes, saves classes from being dull and boring and makes them attractive (Samur, 1983, cited in Karabacak, 1996).

It may be claimed that plays can be used for defeating students' fear of school and class, improving a positive attitude toward a class, developing high-level skills, increasing academic achievement, advancing self-regulation skills and enhancing in-class motivation. Coşkun (2006) states that teaching through plays is a teaching technique enabling to reinforce what is learnt and repeat in more comfortable medium. Furthermore, he explains that plays ensure the most inactive students participate in the activities and create a change in interclass activities by providing a pleasant and comfortable environment in class (Coşkun, 2006).

When school life is taken into consideration, children's problems with math class, their negative attitudes, anxieties and failures are immense in national and international examinations. Among the reasons of this situation, it is observed that there is a change in the reactions of children that can solve mathematical operations from the earlier ages to math class in their school life. According to the researches carried out on those issues, many students around the world show fear and antipathetic response to math class, and many students have difficulties in understanding and learning math (Alkan, 2011). Brush (1979), defending that anxiety for math class increases throughout the time, emphasizes that students' interest in math decreases in progress of time. Hart (1992) states that students have no problems with math, in the initial years of school, or have little issues, but throughout the years passing by, there are declines in their achievement in math.

With respect to the findings of Alkan (2011), it is claimed that the reason why students feel nervous in mathematic lessons is because teachers cannot convey or teach students the mathematical concepts, the cause and effect relation in math or the impact of math in students' real lives sufficiently or enough. Teachers are supposed to explain the purpose of math and its actual relation with real life and bring students in the logic of mathematical operations by including activities to attract students in their classes. On the other hand, in math classes taught via plays, students state that they do not feel they are in math class and their fear fades away. They also indicate that they enjoy participating in classes held via teaching through plays or as if plays, and they do not forget what they have learnt for a length of time.

One of the main problems of Turkish education system is that students are generally anxious about math class, they dislike it or they cannot succeed it. To eliminate this problem, in this study, the impact of teaching through plays in math class of first graders at primary school has been examined. If teaching through plays becomes effective, teachers and students will be taught how to involve plays in math class, the principles of the method and its strengths and weaknesses. In this way, it is thought that beginning from the first grade, negative attitudes, fear and failures would be blocked. Furthermore, it is thought that using the method of teaching through plays in math classes would set an example for future research that will be carried out by presenting theoretical and practical sample plays and activities.

From this point of view, the aim of the study is to examine the effect of educational plays on achievement and retention. Specifically, the study investigates the following four hypotheses and one sub-question:

- H1: There is a significant difference between first graders' pre-test and post-test scores in experimental group where "the unit of natural numbers" is taught through educational plays in math class in primary school.
- H2: There is a significant difference between first graders' pre-test and post-test scores in the control group where "the unit of natural numbers" is not taught through educational plays in math class in primary school.

- H3: There is a significant difference between first graders' achievement in experimental group where "the unit of natural numbers" is taught through educational plays and the control group where "the unit of natural numbers" is not taught through educational plays in math class in primary school.
- H4: There is a significant difference between first graders' retention in experimental group where "the unit of natural numbers" is taught through educational plays and the control group where "the unit of natural numbers" is not taught through educational plays in math class in primary school.

1. Sub-problem: What do students think about teaching math through educational plays?

    a) What game did you have fun playing in math class? What amused you to do in this game?
    b) How is playing games in math class different from games you play at home or outside?
    c) Do you want to learn through playing games or writing in math class? Why?

Chapter 1 Play and Pedagogy of Play

Abstract In this chapter, play, historical development of teaching through plays and the definitions of play are discussed. Moreover, play criteria, play approaches and principles of teaching through plays are tried to be explained.

Keywords: play, play criteria, play approaches, play pedagogy, teaching through plays

Play

Various definitions have been made since play was approved in education. The concept of play may be defined differently on each approach. Lazarus (1883) asserted that play was a free activity bringing joy, emerging spontaneously and without an objective. Groos (1899) considered play as pre-testing for the maturity reached at the end of childhood. For Hall (1906), a child reflects cultural improvement of humanity in his/her plays. While Seashore was defining play as expressing oneself for the sake of the pleasure of expressing oneself, Frobel explained play as the first natural sprout of childhood. According to Dewey, plays are conscious activities without expecting a certain outcome. On the other hand, for Bruner (1976, 31) plays are a special way of breaking stability while Schiller explained them as spending enthusiastic energy aimlessly (Spodek & Saracho, 1987). Bilen (1999) describes play as "activities to improve physical, mental skills of individuals, make life enjoyable, developing artistic and aesthetic qualifications and skills." The definitions indicate that play has a wide-ranging quality serving as a common ground for other disciplines. It can be said that play, which has survived throughout the history, is used for different purposes in educational environments.

Historical Development of Play

All human beings are active seekers of knowledge, and play is an inseparable complement of this continuous pursuit.

The pedagogical significance of play is not because it is a kind of teaching technique with a couple of particularly constituted activities defined as "play," but rather it is considered to be the primary environment for children's learning and improvement. It is predicted that along with the changing needs of information society, individuals of future generation would have more consistent, creative and complex cognitive skills and communication skills. Many of those predictions bring up a more individualized, complex and interactive schooling idea in today's education applied in different disciplines. Approaches recently developed in math, science and reading-writing education are in need of plays that students can constitute their own ideas via fun activities (Bergen, 1998; Bergen, 2009).

When play and child-oriented approaches are examined, it can be said that the first activity-oriented program about early childhood is kindergarten developed by Friedrich Froebel in Germany 150 years ago. Froebel created his education program, choosing required materials for activities after observing children. He defined his educational materials as "gift" and activities as "duty/task." While "gifts" are several objects such as balls, wooden blocks and other materials, "duties" consist of many crafts like cutting, folding and weaving paper. Even though Froebel often mentioned freedom in his articles, his activities are considered to be normative. In fact, since children are responsible for carrying out whatever duty they are entitled to do, it is hard to characterize those activities as plays.

In Italy, 50 years later, Maria Montessori designed a different early childhood education program. She observed children's play and created her activities by placing those observations in a theoretical aspect (Montessori, 1965; 1973). Although kids are told to manipulate the materials in activities, those activities continue as rule-based plays (games). Despite the fact that play observations serve as a source for programs in both approaches mentioned above, it can be said that different theories have been emerged and final education program does not include plays but educational activities. Along with the institutions for preschool education expanding in Western societies beginning from 1900s, plays have been accepted as the legitimate educational activity. Later on, educators have observed children and accepted what children did during the play to be authentic, vital and potential to learn. In their observations, they have seen how children tested their ideas, abstracted information and took action based on that information.

Even though plays have been appreciated and cherished more due to Frobel's educational approach and Montessori's teaching techniques, researches who are considered to be pioneers in education have preceded paying attention to Dewey's educational approaches. Dewey stated that he disagreed with Frobel in terms of Frobel's approach claiming imaginary plays consist of materials detached from reality and a sort of symbolism (abstraction) (Franklin, 2000). In imitation, impersonation plays, Dewey recommends to use materials that can pioneer the development of guidance, experience and expectations in constructing children's ideas, especially realia. As many as natural materials about life are used, structures used as sources for children's ideas will be represented in plays. Advocating that materials such as kitchen utensils and household goods are more realistic, interesting and beneficial. Dewey argues that plays with some mystic materials and rules for adults like in the Five Knights play by Frobelin do not correspond to the reality of children and do not represent their daily lives (Dewey, 1990/1915, p. 123–124, cited in Franklin, 2000).

Dewey considers impersonation plays in the center of mental development. Dewey defends that mental cohesion and complementation of children can be provided by the means of free plays and various materials presented to the children. Stating that plays should have introduction, development and conclusion parts played under the guidance of adults, Dewey states that children can realize their mental development through processes such as providing them the proper environment to display expected behaviors in plays, retention of play process and making them responsible of those activities as a result of their own behaviors (Dewey, 1933; Franklin, 2000; David, Goouch & Powell, 2015).

What Is a Play?

Reviewing and using play together with learning processes date back to ancient times. While studying plays within the scope of various disciplines along with schooling process cause the emergence of different approaches, the fact that many activities are defined and used as plays can be listed among the features to indicate the increasing interest in plays. In this context, Bergen (1998; 2009) introduces five different characteristics of play:

1. Play is a channel and instrument for communication

While language provides social communication for adults, play, as a result of improvement in children's language skills, serves as a primary channel conveying children's opinions and feelings to others. When adults observe children's plays, they encounter misconceptions and deficient concepts of children. This situation helps adults understand the thinking processes and profound feelings of children.

2. Play is the technique and materials with which an artist uses

While adults use various media organs to reflect their artistic and creative products, plays are vessels that children show their initial artistic reflections. As play functions as an instrument to test materials via different points of views, it also enables children to discover how to use objects for different intended purposes or how to use their own bodies. Moreover, children can use play combining with drama, art, music, dance and literature in the pursuit of active information.

3. Play functions as a means, instrument and a communication organ for what has been or is being told

Helping children to turn their experiences into immanent senses and processes, play improves children's perceptions of self-adequacy.

4. Play functions as a mediator just like driving power of air or a strong balancing influence

Through plays, children reinforce their activities and gain deep concentration and attendance. Furthermore, softening life realities, play functions as a filter and stabilizer in children's undesired experiences and confrontations, and helps children to have those experiences in risk-free environments.

5. Play functions as a natural habitat for children

Plays help children improve natural habitat to provide a rich learning environment regarding their social, emotional, cognitive and physical progress. Natural habitat is the environment where a living being grows best, and natural habitat for children is play.

Each definition indicates that play has different interpretations and consequences. Play is different from "duty/task." In this sense, researchers have developed different approaches and principles regarding whether an activity is a play or work. Hence, it is stated that play criteria cannot be observed. According to a particular approach, satisfaction from an activity determines whether an activity is a play or not. If activity is solely meant for its own sake, it is a play; if it is carried out for an external reward, money or gain, then it is a work. Besides, seriousness of the activity is an important quality to separate play from work. Activities considered to be light and unimportant may be thought as a play (Spodek & Saracho, 1987, Fisher et al., 2011). In this context, knowing the criteria of the play can be beneficial for us to gain profound knowledge about the concept of play.

Criteria of Plays

Sources recommend several criteria for plays (Lieberman, 1965; Neumann, 19741; Rubin, Fein &Vandenberg, 1983; Spodek, Saracho & Davis, 1987). Rubin et al. (1983) summarize six criteria to define play, such as 1) inner motivation 2) aiming meaning rather than result 3) internal control rather than external one 4) activities not without but with instruments and means 5) freedom instead of imposed rules and 6) active attendance or involvement. However, Neumann (1971) mentions three criteria, which continue through a line from work to play:

1. **Control:** There are differences between internal and external controls in activities. If control is internal and shared with a friend, activity is a play; if the activity is under the influence control, it is called a task.
2. **Reality:** There are differences between internal and external realities. Activities that suspend reality, pretend, imitate and imagine and suppressing external reality are recognized as plays, and plays strictly related to life and external reality are determined as plays.
3. **Motivation:** Activities through inner motivation may be considered plays.

Rubin et al. (1983) describe play based on different criteria such as objectives, materials, rules and other elements in a play.

1. Plays are activities that are not directed by desires, impulses and basic needs, but motivated individually by contentment and gratification immanent in an activity by itself.
2. Players are interested in activities rather than objectives. Objectives are designated individually, and behaviors of players are spontaneous.
3. Play begins with known objects and continues with the discovery of unknown objects.
4. Plays can be formed of figurative (unreal) objects.
5. Play is played in active attendance of players (Spodek, Saracho and Davis, 1987).

About the qualifications of play, Rubin (1998) states that:

- Play is not a repetition of a play, being manipulated by others and things done for a reward; it is carried out via inner motivation.
- Play is not target oriented; it has its own satisfaction peculiar to itself.
- Play is not directed by rules. It is different from games with contests and rules.
- In a play, children attribute meaning to objects themselves; they raise the question "what can I do with this object?" rather than the existing meaning of them.
- Play includes figurative elements of which relation with reality is not strict. Objects and activities acquire new meanings, objects are perceived in a different sense from their old qualities and context and transformed into new ones (Rubin, 1998).

In order to test experimentally which criteria is used most for designing activities as plays, Smith and Vollstedt (1985) decided on the criteria as follows: inner motivation, metaphors, positive impact, flexibility and discernment of cause/effect (without expecting an end/result, with an immanent satisfaction). While observants mostly preferred metaphor, positive impact and flexibility as the most useful play criteria, many observants excluded inner motivation in the evaluation. At the end of the research, it is concluded that entertainment, flexibility and impersonations are criteria to determine a play. To sum up, plays are enjoyable behaviors internally motivated, consisted of impersonations and imitations, not directed by rules and not target oriented (Rubin, 1998).

Play Approaches

Determining certain criteria and definitions cannot help to understand why people and especially children play games, this continues as one of the difficulties that psychologists have been facing for about a hundred years. Various theories have been suggested about why people play games. In terms of two theories on plays, one of which is classical and the other is modern, Mellon (1994) summarized why people play games. According to Gilmore (1971), classical play theories suggested before 1920 tried to explain the existence, meaning and biological and hereditary viewpoint of play using evolutional and psychological concepts about. Classical play theories were listed as:

- **Surplus Energy Theory** by Friedrich Schiller (1878),
- **Relaxation Theory** by Patrick (1916),
- **Recapitulation Theory** by Stanley Hall (1906),
- **Pre-Exercise Theory** by Karl Groos (1899).

When it comes to the modern play theories, they focus on the psychological value of play and its importance in a child's social, cognitive and emotional development. Unlike the classical play theory, modern play theory which is supported by experimental researches emphasizes high level and symbolic thinking. Modern play theories can be listed as:

- **Psychoanalytic Play Theory** by Freud (1923/1973),
- **Arousal or change of pleasure theory** by Berlyne (1969),
- **Metacommunicative play theory** by Bateson (1972) and Frost (2010).
- **Cognitive play theory** by Piaget (1952) and Vygotsky (1967) (Biddle et al., 2013).

It is evident that there are different points of view in respect with classifying theories as well as defining plays. Ferholt (2007) categorized play theories under five different titles:

(1) Biological play theories (Schiller and Spencer; Lazarus (1883); Emler and Mitehel; Groos),
(2) Psychoanalytic play theories (S.Freud, A.Freud, Klein, Erikson and Winnicott),
(3) Cognitive play theories (Piaget, Vygotsky),
(4) Intercultural psychological play approaches (Göncü & Gaskins),
(5) Anthropological play theories (Hall, 1906; Turner, 1969; Huizinga, 1970, Bateson, 1972, Geertz, 1973 and Schwartzmann, 1978) (Ferholt, 2007; Biddle et al., 2013).

A. Classical Play Theories

1. Surplus Energy Theory

In Friedrich Schiller's (1878) theory, play is considered to be the method to express extra/surplus energy existent in all living beings after fulfilling the biological basic needs. Play is defined as an activity to reloading or refreshing the energy spent by individuals. Play, which is thought as the opposite of work or task, is considered to be an exercise of renewal, whereas task is recognized as unpleasant responsibilities and burdens.

2. Relaxation Theory

In Patrick's (1916) theory, unlike the Surplus Energy theory, individuals regain energy spent in the play and revive. He defends that after working for a while, individuals should rest and restore energy for the work, in other words, they need plays.

3. Recapitulation Theory

Stanley Hall (1906), in his play theory inspired by Charles Darwin's evolution theory, states that children trace their ancestors from primitive skills towards modern skills in plays, which he considers as a repetition of the progressive process of human race.

4. Pre-Exercise Theory

Karl Groos (1899) examined first behaviors during a play first in animals and then in human beings. He observed behaviors, traditions and competitions of adults in children's many plays. According to Groos, play includes behaviors similar to the roles of adults. The main purpose of a play is to make children exercise the experiences to prepare them for their future life.

B. Modern Play Theories

1. Psychoanalytic Play Theory

Freud advocates that play bear a great importance in emotional development of a child. Play has a therapy impact providing a change in emotions from negative to positive. Through this treatment, children prevent themselves from traumatic experiences by having the liberty to act freely. Moreover, plays enable to notice stressful activities and be influenced less by their consequences or incline to more positive behaviors (Biddle et al., 2013).

2. Arousal (stimulation)/Change of Pleasure Theory

While defining his theory, Berlyne (1969) built his theory on the fact that how plays can provide children to find beneficial information through different emotions in order to learn what is happening around. His theory explains how familiar and different impulses function in the discoveries of children. Different impulses attract children's attention whereas similar and ordinary impulses can cause boredom.

3. Meta-Communicative Play Theory

Bateson (1972) and Frost (2010) claim that children's plays are pretentions, namely imitations or impersonations and they imitate real life behaviors in that process. Children play an imaginary play constraining reality with an object or convert a real life behavior into a play or something close to reality. In this theory, play includes meta-communicative points of view in terms of what two individuals perceive in cultural and individual reality.

4. Cognitive Play Theory

Cognitive play theory of which masterminds are Piaget (1952) and Vygotsky (1967) can be summarized as cognitive construction and process of thinking. According to Piaget, play is an orientation period. Play is a method that a child, by his/her own experience, can learn the subjects and skills nobody can teach him/her. Piaget defines three successive systems –training play, symbolic play and play with rules (games) – and outlines the development of children's plays in the first seven years of their lives (Nicolopoulou, 1993). For Piaget, children gain information in two different processes, namely assimilation and accommodation. In assimilation, children get information about materials around them whereas in accommodation, they create a situation of balance and unbalance by matching their preset and new information. Piaget claims that while assimilation and accommodation happen at the same time, new concepts and opinions emerge in-between those two processes. The process of assimilation takes place predominant over accommodation in plays. Children tend to assimilate new materials and opinions in plays (Biddle et al., 2013).

Piaget (1962) defines play as the process when external structures are abstracted and manipulated and then made to fit one's own organization schema. While play enables to improve children's intelligence, it continues to have influence on children's mental behaviors as they grow up. Piaget mentions three different steps related to the development of play. The first one is **sensorymotor period** babies have, based on imitated and physical behaviors; the second is **symbolic play**, which is the period of imitative and dramatic play reflecting characteristic features of preschool children and the third is **the period of play (with rules)**, which can be called typical childhood plays. Children are more prepared for and tend to play throughout early childhood and when they enroll in primary school. Dramatic and symbolic plays can be assumed as a describing style. Once children depict the outer world mentally, they can manipulate those elements of world by using assimilation and accommodation processes (Spodek & Saracho, 1987). Since the process of assimilation in a symbolic play permits a child to change reality in his/her own way without having the real-life oppressions, it is more dominant and overpowering with respect to the process of accommodation. In the first period, namely sensory-motor play, the process of accommodation can be said to be overpowering (Saracho, 1986).

Fun in a symbolic play stems from children's ability to change and manipulate meanings (Piaget, 1962). Children develop several aimless behaviors. Children recognize plays as a way of enjoying oneself, and at the same time, adaptation period to the hard times when they overcome new concepts, skills and emotions. Thus, plays can be thought as a unique challenge and opportunities as a result of that challenge developed by children themselves.

According to Vygotsky, real play starts at the age of 3 with pretentious plays (as if plays) which children do not discern from socio-dramatic play. Being a constant social symbolic activity, play also enables children to form their way of comprehending sociocultural materials of their own society and use them with the intention of play (Nicolopoulou, 1993). Vygotsky (1967) believes that conflicts and problem solving are the fundamental characteristics of development. The focal point of studies is the idea that learning is realized through social interactions. He explains his theory in three sociocognitive processes:

1. **The zone of proximal development** indicates the difference between what students can achieve under the guidance of teachers, peers or parents and what they can do on their own.
2. The fact that interpersonal knowledge turns into individual knowledge is caused as a result of understanding the way of constructing concept improvement carried out by **interiorized speeches** with two or more people's interactions.
3. Transition to explicit and clear rules from implicit rules is based on the idea that children **remember the roles they take and also processes of playing friendly during plays**.

According to Vygotsky (1967), while it is stated that many mental skills are developed in dramatic and symbolic plays, pretentious (imitative) behaviors enable to understand the objects depicted in dramatic plays much better. Vygotsky (1967, 1978) claims that using those objects support the development of students' opinions (Biddle et al., 2013).

Play assists children to learn their own world as well as it helps them to explain their feelings and opinions and improve their social relations with their peers. Furthermore, play is also beneficial for children to acquire knowledge to develop new ideas; compare them; discover the contrasts relating to their old ideas; improve acceptance, refusal or opinions and change them. Educational programs considering children to be active learners include processes that provide experiences to organize and improve their own knowledge. The educational function of play is related to cognitive, creative, linguistic, social and physical development (Saracho, 1986).

Types of Plays

Along with different play approaches in the literature, various kinds of plays are described by researches. These plays can be categorized as cognitive, social, creative and manipulative plays. In the literature, researchers' perspective of play varies in categorizing plays, too.

Cognitive Plays

In a cognitive play, children can create objects and roles. Even though they use an object that is a representative of something, they act loyally in terms of the definition and purpose of object. Children are aware of the use of objects and roles both in their real function and unreal, imaginary, out of purpose handling. Smilansky and Shefatya (1990) classified five different plays in the light of Piaget's cognitive play approach:

- **Cognitive plays** (sand play),
- **Construction plays** (cubes and block plays),
- **Dramatic plays** (firefighter and cops and robbers plays),
- **Socio-dramatic plays** (pretending to be a doctor and playing house),
- **Plays with rules** (sportive plays).

Creative Plays

Creativity can be defined as the process of discovery occurred outside the imitation, restrained by the process of imagination. Creative individuals ignore what is common and ordinary in order to produce new, authentic and unique one. Meanwhile, they involve in play processes to use objects and incidents around for different purposes and functions.

Social Plays

Socializing process requires a child to get on well with others. In social plays, children improve verbal and non-verbal communication skills related to their peers' attitude, feelings and opinions. While playing with their peers, they discover that their peers' points of view are different than what they think. They either assimilate those thoughts or recognize an acceptable level of adaptation of that point of view or social atmosphere to accept it. While oral resources are taught thanks to sharing and coordination, social and cognitive development of children can be provided. Explaining social developments of children in plays, Parten (1932) classifies plays under six steps:

1. Unoccupied play
2. Solitary play
3. Onlooker play
4. Parallel play
5. Associative play
6. Cooperative play

Manipulative Plays

Play materials enable children to participate various play processes. Play materials support both narrative plays such as imitation, demonstration and non-verbal, manipulative, structural, object plays. As well as there are particular manipulatives used in subjects peculiar to math class (Dienes sticks, Cusienaire sticks, ten-cubes, geometric shapes...), materials with different qualities are also handled in this context.

Apart from those plays, it is possible to see many play types. It is seen that plays are classified based on the purpose of use, materials used, developmental period or actions taken. In this sense, Hughes (2002) mentions 16 different play types:

1. Symbolic Play (riding a broom, flying with wide arms);
2. Rough and Tumble Play (rolling, tumbling);
3. Socio-Dramatic Play (impersonation of parents, cooking);
4. Social Play (plays imitating social rules, dialogues);
5. Creative Play (changing objects or events, transformation, establishment of new connections);
6. Communication Play (jokes, gestures, sound similarities, charade);
7. Dramatic Play (Preparing a TV Show, imitating an incident on the street);
8. Locomotor Play (skipping rope, sportive plays);
9. Deep Play (climbing a tree, jumping high, playing with matches);
10. Exploratory Play (plays to discover qualities of objects, catching, throwing, tasting);
11. Fantasy Play (There is no rule. Flying, impersonation of a pilot-captain, driver);
12. Imaginative Play (There are rules. Realistic. Imagining an absent object as if it were present);
13. Mastery Play (drilling, constructing waterways, playing with sand);
14. Object Play (playing with materials such as dress, glass, spoon out of their indented purpose);
15. Role Play;
16. Recapitulative Play (dramatizing primitive times, past, ancestors and historical incidents).

Since plays we encounter in different types almost have the nature bundling all of our behaviors up, it does not seem easy to stereotype them. When it comes to using plays as a teaching medium/environment or method, it is crucial to pay attention to different disciplines such as developmental psychology, sociology, economy and field to be applied (class, subject and discipline). An educational program that uses principles of teaching through plays sufficiently can provide an educational atmosphere where the unique developmental environment of play can be felt at the utmost level.

Principles of Developing Pedagogy of Play

Children begin to understand the world around them in their plays. Observing plays, teachers can understand how students comprehend the world and form opinions on how to make learning easier. Teachers should make learning easier by not being moderator or leader but taking facilitative roles. Teachers should be well equipped about developmental and educational results of play programs. In a well-designed program, teachers should be able to use processes to transfer social roles and the world of adults to learning. Even though many programs use plays as teaching means, in order to attain more influential consequences, teachers are required:

a) to provide and present a play medium supported by sufficient material, equipment and playing field,
b) to nourish positive social communication,
c) to improve creative play, and support it,
d) to learn how to use various types of plays efficiently regarding children's needs and interests,
e) to take active roles in beginning and continuing plays (Spodek, Saracho and Davis, 1987).

To attain influential consequences in plays, paying attention to play principles makes play process more productive. Those principles are reviewed in the context of starting and expanding play, arranging physical and social environment in plays, and in terms of play, learning and teaching programs.

1. Beginning a Play

Plays generally begin when children are supplied play materials. Those materials should be attractive; they should provide proper playing opportunities and present some innovations. Planning can be carried out about what to do during a play. In this planning process, preliminary processes are required to be reviewed, namely introduction of new materials and toys, purposes of materials and usage and limitations of materials.

2. Expanding a Play

Teachers feel the need to guide students by expanding their games or intervening to make plays more productive and assisting students. Even though most of the time they do this by involving in games, they are not supposed to take moderate roles to support children's being free. Avoiding interferes to make a play as a process of following the rules and instructions, they should prevent children from distracting, make them feel to be in the center of the play, and also abstain from any kinds of behaviors that will prevent students from taking individual responsibilities.

Besides, to make a play continue and be productive, teachers can benefit from play at top- level by:

a) Adding materials to or taking materials from play field,
b) Advising new roles,
c) Changing setting (place), friend and partner when children are obstructed.
d) Speeding up and slowing down time (play).

In this sense, Hughes (1996) mentions that enriched play environment can provide opportunities to expand the play. Those enrichments can be listed as:

- Physical environment with interesting and different qualities (places with different sizes, objects, tools and things);
- Physical environment that can be intervened and changed (sportive plays, chess and cube plays);
- Opportunities to play with natural materials (water, air and dirt; snowball, kite and digging plays);
- Opportunity to act (running, jumping, rolling and balancing);
- Using natural and artificial materials (using art or cooking equipment);
- Activities to stimulate sense organs (music, using different scents and colors, using food and beverages, using soft and hard objects);
- Observing natural and artificial changes around (seasonal changes, building exercises such as construction, demolition or painting);
- Social interactions (communicating with different ages and groups and deciding to work alone or in a group);
- Taking role or impersonation (role-playing with clothes and costumes and taking responsibilities);
- Experiencing different emotions (defining places one likes, dislikes, hates and gets bored of and dramatizing) (Hughes & Melville, 1996).

Teachers, parents and therapists intervene to a certain extend with children's behaviors and developments. Those interventions may be explicit or implicit, direct or indirect. No matter how children are intervened, people in charge are required to do this consciously. If the intervention is conscious, this process should be fine, careful, evaluable and improvable. Meanwhile, the main issue may be the fact that whether plays and intervention can continue at the same time or not. Neumann (1971) advocates that activities can be determined to be played as long as teachers have control over children's activity (play), determine the level of reality and provide motivation for the play. Ellis (1973) defends that plays can emerge as a result of answering some questions. Those questions are listed as:

a) Whether a child is expecting external or experimental rewards;
b) The fact that whether child's behaviors are directed externally or self-directed;
c) Whether children are forced to recognize reality and limitations (roles, duties, responsibilities and rules) or those limitations are suspended or not;
d) Whether relations between incidents and consequences in implementations are imposed or relations are set free.

The questions above may be used as a mentor to determine whether children study (work) or play in activities.

Teachers can begin new activities using children's plays, their present play patterns, without forcing them, but inviting them into small groups. Those who do not willing to join a play or group activity should be tolerated and directed to new activities. Play activities require children to be ambitious by considering others' feelings, with cooperation, being easygoing and without being aggressive.

Teachers have developed various play intervention strategies to make children's plays educationally productive. Those may be direct teaching, praise-reward, protecting-taking precaution, communication, display-demonstration, directing and attendance (Spidell, 1985). The fact that play is hard is because adults consider plays as hard processes in need of intervention. Trying to understand why a child plays and its consequences will be helpful for educators to perceive play as a potential rather than being a hardship (Spodek & Saracho, 1987).

3. Arranging Physical and Social Environment in Plays

Efficient teachers use the method to create an environment where teaching through plays can be implemented and opportunities to increase educational value of play are provided (Spodek & Saracho, 1987). According to Piaget (1952), there are four interaction processes to support mental development: 1) maturity, 2) physical interaction with objects around, 3) social information transfer emerged as a result of experiences and human interactions and 4) balancing process (child's own effort to construct information). In two of these processes, educators have direct influence. They are physical and social environment of children. Both of them affect children's improvement in play. In addition, physical environment includes sensory elements such as sound, scent, taste, texture and also material types, equipment, quality and quantity of play field provided and time spent. Also, physical environment has qualifications such as form, potential, natural environment and climate of school. Social environment also consists of elements such as quality, frequency and feature of adult and peer interactions realized by children in plays; regional, cultural and national values and traditions and customs. Taking these elements into consideration and their interaction with each other will positively support educators who organize a play. In this context, educators are required to pay attention to principles of arranging environment in order to comprehend children's play and learning improvement and factors in physical and social environments, and plan effective play fields (Bergen, 1998). In this sense, principles of arranging environment are tried to be explained via the concept "load" by Mehrabian (1976). According to Mehrabian, "load" means density of stimulus around, information transfer ratio and quality of information. While some factors constitute high-loaded environments (enhanced information environment), some are called low-loaded environments. "Load" is used for explaining innovations and complex structures around. Innovation is related to the level of uncertainty and obscurity level around and is the indicator of what would happen in the environment. Complexity means qualities, sorts and variabilities of elements in an environment. Children and adult have behaviors of getting closer to or avoiding environment. While discovering is a behavior of attachment and affiliation with a high performance, not discovering is called neglect, boredom, alienation and avoiding with a low performance. Different environments loaded high and low may cause different behaviors. Thus, Mehrabian (1976, p. 12–13) summarized the characteristics of high- and low-loaded environments (see Table 1).

In this context, while environment is being arranged regarding the play, high-loaded environments can be preferred. Moreover, security precautions related to the environment, rules such as convenience with age and development of children can present qualities to increase the quality of play. It can be said that children can display much more behaviors peculiar to play in environments where they do not feel the social roles and oppression much, and

Principles of Developing Pedagogy of Play

Tab. 1: Evaluating High- and Low-Loaded Environments

High-Loaded Environments	Low-Loaded Environments
Uncertain	Certain
Various	Boring, similar
Complex	Simple
Novel, not encountered before	Familiar
Large scale, long-termed	Small-scale
Includes conflicts	Includes similarities
Intermittent	Continuous
Surprising	usual, ordinary
Heterogeneous	Homogeneous
Crowded	Calm, dull
Active	Steady, no action
Rare, unique	Ordinary, common
Random	Patterned
Impossible	Possible

Source: Mehrabian, 1976, p. 12-13

rules were determined before. Observing children's behaviors of play, the best environmental arrangements can be done exclusively for individual, class and groups.

About features of play fields used by children, it is stated that new and complex environments reveal exploratory behaviors rather than play behaviors in the research carried out about when children explore something and when they play. It is emphasized that average new and complex environments reveal behavior of play (Ellis, 1979; Hutt, 1979; cited in Bergen, 1998). Play area is required to include arrangements to pay attention to developmental qualities, individual differences and cultural and social structures of children to use play as an educational instrument. If the arrangements are handled according to the principles of environmental arrangements, sufficient innovation, complex environments, and high-level benefits from play will be attained (Bergen, 1998).

4. Exploration, Play, Learning and Teaching Programs

Defining so many activities as plays requires educators, who demand planning a learning environment for children through plays, to be sophisticated and equipped regarding the features of the play. In choosing influential play

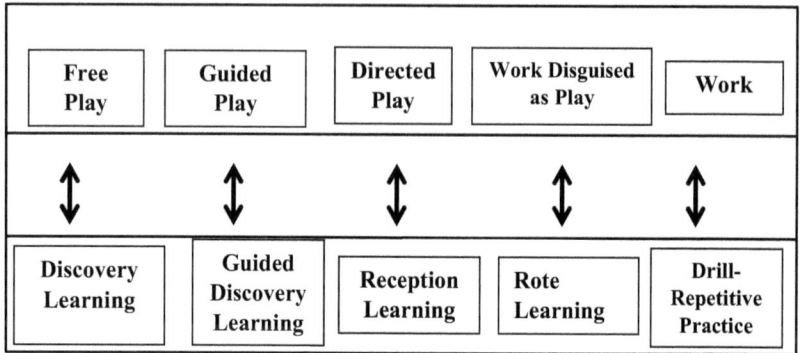

Fig. 1: The Schema of Play and Learning
Resource: Bergen, 1998

experiences to make children reach their learning objectives, relations and distinctions between learning and play can function as a guide for educators.

Researchers and theoreticians define play as an assimilation process rather than cognitive accommodation processes, started by child himself/herself. Fein & Schwartz (1982) explain that assimilation process is dominant over accommodation process in a play, hardship and challenge is neither presented by environment nor desired in the environment, but created by child himself/herself. Similarly, Hutt (1971) explains the difference between exploration and play as follows: "in exploration, children examine what objects are able to do, whereas in a play, they discover what they can do using those objects."

Schema related to learning and play consists of behaviors changing between two poles, namely free play, which is started by child and includes high-level internal control components, and duty-task, which is started by adult (someone else other than child) and includes external control components. Behaviors including assimilation component are arranged as the behaviors on the far left (internal) whereas behaviors including accommodation components are arranged as the ones on the far right (external). Processes that are plays and not plays are named under the changing categories from free play to work. Learning process is listed from exploratory active learning to practice and repetition activities (Bergen, 1998).

In **free play,** there is high-level inner control, reality and motivation. Player decides what to play, when and how to play himself/herself or whether he/she plays or not. Moreover, he/she decides whether play is individual or with partners, and who the partner will be. Again, it is the player who determines environment arrangement and medium. Play can be performed in or out of the class with limitations such as time and medium used. In free play where processes of **exploratory learning** (active learning) are dense and profound, there is spontaneous manipulation of objects around and social interactions with adults and peers. In this context, regarding exploratory learning, Dewey mentions interaction which enables children to test the results of their behaviors, whereas Bruner and Piaget stress that cognitive development is possible through exploratory learning processes where children will improve problem-solving skills and get feedback as a result of their interaction with environment. On the other hand, Vygotsky advocates that children should learn with social interactions and exploratory learning, with adults and peers, using opportunities provided by cultural environment, whereas Dienes mentions that discovering objects through free play, children carry out abstractions and mathematical operations (Dewey, 1933; Piaget, 1962; Bruner, 1966; Dienes, 1967; Vygotsky, 1967).

Guided plays are plays with social rules, certain roles, some realities, motivations and control should be paid attention by children. Just like in free play, children have a wide range of games; however, there are social rules, motor behaviors, security and sharing. Related to using environment, children are not as free as they are in a free play. Plays with materials, play tables and stations prepared before can be mentioned as examples. During a play, while adults are leading players to individual or group plays, players can set rules for each other. Free play arrangements in preschool can be examined in this group (Weisberg, Hirsh-Pasek &Golinkoff, 2013).

In guided plays, active learning processes can be studied in the context of adults' certain objectives. In guided plays generally performed in educational environments, social and cognitive interactions help learners define the subject better and learn by taking as an example under the guidance of adults. Walking around during a play guides lead players about materials or rules.

Directed plays include many external factors and instructions determined by adults, and they are generally performed by the consent of adults. Children can reach inner control, reality (as if...) and possibilities to experience motivation opportunities in a way chosen, created and directed by adults. Directed plays can be enrolled in this category. Children do not have the option or freedom to decide what, when and how to play. Many arrangements are made before from choice of play, to whether it is a group or individual play, from game rules to time. This type of play is generally considered to have more and profound accommodation. Playing in environments where children can have assimilations, namely new explorations, where they meet new schemas, where they do not wait for others to finish the play or where they are not restricted with time and space can enable them to construe those activities as enjoyable, attractive and interesting plays.

Plays directed by **meaningful learning** of Ausubel (1968) are generally considered directed plays. In this verbal explanation-dominated learning, children are asked to make verbal explanations by presenting first-hand experiences. Learning environments where verbal learning, cognitive structures and essential arrangements are provided properly are considered to be meaningful/expressive. According to Ausubel, this process has similarities with the qualities of concrete operational stage by Piaget at the ages of 6–7. After educators evaluate plays and experiences of students, they are required to organize activities with necessary arrangements, rules and materials.

Work defined as play and task (work as if play) that are duty-oriented unenjoyable activities can be transformed into guided or directed play by the means of inner control, motivation and reality arrangements. Singing about any subject, trying tongue twisters, asking riddles and puzzles, contest of mental-adding up are duties to be defined as plays. In the activities as if games where verbal and motor learning occur as rote learning, players carry out activities by memorizing or repeating rules and behaviors. While restrictions in this game are time, space, rules, partners and groups, meaningful learning cannot be said to be realized since children are provided meaning chains piece by piece.

Work can be defined as activities performed and directed by external motivation in order to attain a determined objective. There is no option to demolish reality (rules) related to task. It can be seen as processes that require repeating accommodations, namely schemas learnt.

Children turn duties into plays in the education process somehow, and while doing so, it is seen that they use principles of reality, control and motivation to extend, to push the boundaries of time and space. This situation may be described as undesired behaviors by teachers at school and parents at home. Children can learn at an early age that activities considered to be tasks can be enjoyable and satisfying. Because being obliged to study hard, they may encounter unwanted processes throughout their lives. In the learning without meaningful, motivation is generally provided by external rewards and punishments. As an example of this kind of tasks, memorizing the multiplication without comprehending or memorizing vocabulary in language learning can be mentioned (Bergen, 1998).

Rosberg (2003) evaluates children's experiences in the light of different criteria and work experiences. This evaluation is given in Table 2.

Even though many educators think that only free plays are real plays, there are educators who use directed and guided plays in preschools. While arranging

Principles of Developing Pedagogy of Play 43

Tab. 2: Evaluating Children's Experiences in Terms of Play and Work

	Play	Work
Energy Level Used	low	high
Clarity of Objectives	undefined	defined
External Symbols of Evaluation	not necessary	necessary
Types of Skills Used	various	few and arranged
Satisfaction Gained	much and frequent	periodical
Suspension of Judgment	often	seldom

Resource: Rosberg, 2003:7

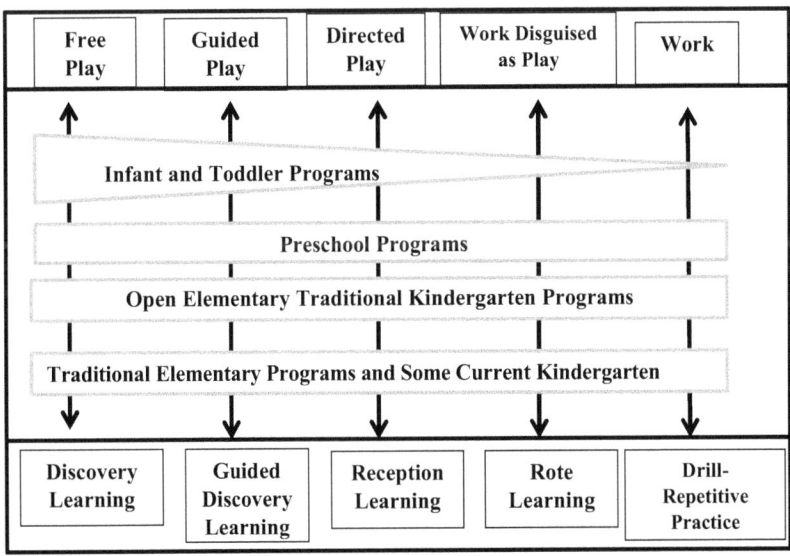

Fig. 2: Usage of Play and Learning Varieties in Different Education Levels
Resource: Bergen, 1998:118

and organizing education environment/medium and programs, educators tend to change plays from free plays to "duty/task" plays according to children's age. Educators' arrangements in programs and environment can be based on providing balance of play and learning in learning environment. The use and distribution of plays in educational environment in four different levels of education is showed in Figure 2 (Bergen, 1998).

As indicated in Figure 2, free play begins from infancy, and it is used in all of the educational programs; however, its influence gradually decreases whereas learning types based on memorizing and exercises in the context of work-oriented activities become more significant in primary school programs.

In traditional primary school programs, it is seen that free plays are less preferred in comparison with guided games whereas plays and learning preferences in programs of preschools and enriched programs of primary schools are equally distributed. Learning types based on memorizing and repetitions that increase when children start school can be enriched by the means of enjoyable and instructive atmosphere of plays, and thus programs can be designed to advance students' development.

Characteristics of Teaching through Plays

In addition to its importance in every step of life, plays are considered a unique learning environment especially in childhood when physical, emotional, cognitive, social, psychomotor and psychological developments are crucial. Whereas children organize their own habitat by using their imagination in plays, what is learnt by questions and trial and error is reinforced in mind and thus diagrammatized. Children improve abstract thinking skill via plays. Children who deduce possible consequences in setting and narration of reflect many of those consequences attained in their real lives (Uğurel, 2003). Considering the reasons to prefer play as a learning means, play has multiple educational influences, besides, it educates physically, socially, emotionally and cognitively (Moyles, 1989; Wood & Attfield, 2005; Bergen, 2009; Hunter & Walsh, 2014). While play provides renewal of culture and friendship culture established among children, it also protects children from anxiety and devastating thoughts. Moreover, plays provide opportunities for adults who want to have information about the child and his/her needs in order to present children and teachers mutual learning opportunities, create real situations in educational processes (Moyles, 1989). Learning through play can be thought as making children acquire skills to teach them various strategies such as problem solving, reasoning, estimation and conceptualizing (Bransford et al., 2000).

Another reason for learning through play may be the usage of all body and abstraction in play and learning process. Abstraction can be defined as combining many physical activities with high-level cognitive activities such as thinking, reasoning, estimation, perception and reflection. Physical activities are applied in education programs to make students feel well and increase their academic success (DuBose et al., 2008).

Even though it is hard to consider learning as the main purpose of a play, theoreticians have tried to re-explain discussions around play by focusing on the facts that children have right to play, and activities pioneering developmental outcomes. Research on preschool children state that play has a significant role on a child's life and development, and free plays, featured activities such as construction plays and imitation plays develop children's cognitive, social skills and creativity (Lehrer et al., 2014). Besides, Lehrer et al. (2014), who state that there is a limited number of studies on school children's plays, declare that time spent for play by children decreases dramatically when they start first grade; plays are replaced with academic success-oriented subjects and activities, and this restriction may have impacts preventing creativity and causing a disabled identity for children in the long term (Hartmann & Rollett, 1994; Lehrer et al., 2014; Ramani & Eason, 2015). Increased responsibilities, school's own social structure, class atmosphere, teacher's and parents' academic expectations for student may have negative impact on student's development. Math anxiety specifically beginning from the first grade in Turkey seems to be the main problem for students and families to overcome in further years. Negative attitude, gender difference, working memory capacity of students, anxiety of teacher, teaching method of teacher and bad experiences of students related to math can be determined as factors to create a math anxiety (Meier, 2015). Studies revealed that when math is taught through a normative and rule-based approach only seeking to reach an end rather than through a flexible approach by the help of relationships, patterns, conceptual network, math anxiety emerges (Geist, 2010; Van de Walle et al., 2011; Finlayson, 2014). In addition to that, it is stated that childhood experiences lie behind math anxiety of adults (Geist, 2010), and students' future mathematical skills are related with mathematical skills in early childhood (Jordan et al., 2007; Aunola et al., 2004). In long-term research on academic success, it is stated that mathematical and numeration skills one has at the time of starting school are strong identifiers of future academic skills, and it is emphasized that the ability of children to use those skills in mathematical problems is a significant identifier for economic and social success of children in their future (Duncan et al., 2007; Lyons & Ansari, 2015). It can be claimed that overcoming math anxiety beginning from early ages can affect students' attitude, interest and nervousness in a positive way, increase success in math and thus children's future economic and social successes will increase, too. Teaching through plays as a method to teach math can be used as an influential method to minimize student's anxiety.

Chapter 2 Play and Mathematics Instruction

Abstract: In this chapter, opinions as the baseline for teaching math through plays and general features of teaching math through plays are elaborated. And the final part includes responsibilities of teachers regarding teaching through plays along with strengths and weaknesses of teaching through plays.

Keywords: teaching mathematics, play-based learning, types of math play, teacher responsibilities

Teaching Mathematics through Plays

Math can be defined as cultural activities, which have emerged in the cultural history of humanity, passed through rich and important cultural and historical developments and concluded as a multi-dimensional and multi-functional discipline. It can be said that improvement of mathematical thinking skills depends on the reconstruction of symbolic means (vessels) required for problem-solving skill and intentional and persistent-trailed self-reflections of students' own behaviors (Van Oers, 2010).

Beginning from the birth, all cultures develop physical environments including many actions and objects to support math teaching in daily life (Ginsburg & Seo, 1999). Daily mathematics is formed in the framework of Dewey's idea (1976) about raw reactions of children in regards with counting, measuring and rhythmic gradation of things (length, width and size) or the idea of Vygotsky (1978) that children are involved and interested in operations such as adding, subtracting, adding-up, division and defining magnitudes far before starting school and eventually children may have preschool arithmetical skills (Ginsburg et al., 2008). The initial mathematical thinking skills of children improve through spontaneous activities in their own culture and via the emergence of mathematical sense (of symbol, schema and operation). It can be said that those activities improve by the means of cooperative problem-solving processes with more experienced children and adults (Van Oers, 2010). Play, which has an important role in acquiring mathematical concepts and improvement of mathematical thinking and skills, is not associated to math at the first stage. While play reminds of activities to have fun, math is considered to be a subject to be taught in an iron discipline. However, neither play is only an occupation amusing one in infancy and childhood nor math is a whole of concepts to be taught through formal studies and textbooks (Taşkın, 2013).

Mathematical plays can take place intentionally or unintentionally in children's daily life, games, school life and many places they have fun in the context of different themes and subjects.

Through plays, children can learn many concepts such as shape, color, size, magnitude, weight, volume, measuring, counting, place, distance and space; and many mental operations like categorizing, grading, matching, analysis, synthesis and problem solving. Children can learn various mathematical concepts in their plays by manipulating objects, creating groups with different objects and counting those groups (Akman, 2002; Taşkın, 2013). Ginsburg et al., (2008) defends that mathematical plays can be interpreted in three different ways. They are **mathematics embedded in play, play centering on mathematics and play with the mathematics that has been taught in school.**

1. Mathematics Embedded in Play

Math intertwines with children's plays in early childhood just like in other periods. Mathematical plays in which "mathematics embedded in play," includes daily mathematics. Daily mathematics used by children is vital and exciting. Children can develop mathematical strategies, be involved in important mathematical thoughts, use math in their games and play with math itself. Children generally enjoy mathematical works or plays (Ginsburg et al., 2006).

Daily mathematics used by children can be said to be instant, everywhere, sometimes sufficient and sometimes more complex than estimated. It involves processes of different estimations, reasoning and problem solving such as which one is more, which number is greater (Ginsburg et al., 2006). About mathematical skills of younger children and roles of adults, Van Oers states that:

1. "Scratches" of children mean number, quality and quantity change (adding/subtracting) for children,
2. Children get to know communication intention related to quantitative concept using those signs,
3. Children develop those signs mostly via communication,
4. Role of adults is very important for the sake of pronouncing those images shared and defining common mathematical concept (Van Oers & Duijkers, 2013).

In addition to scratches, paintings and communication with adults, plays reflect children's point of view on daily mathematics. Children enjoy using daily mathematics in their plays and even play math learnt at school spontaneously and naturally (Ginsburg et al., 2006). Children can use undirected (not rule-based) informal skills and opinions about numbers, shapes and models while reading or playing with blocks (Ginsburg et al., 2006). Even though play is considered to be essential for cultural and mental development and especially for math, it is not thought to be sufficient since plays help children not mathematicize but interpret children's experiences clearly in mathematical forms and perceive the relationship between these two (Ginsburg et al., 2008). In other words, play does not guarantee mathematical improvement but provides prosperous opportunities. The most important benefits are to enable children to represent mathematical thought emerged in plays and to provide children teacher's support (Ginsburg et al., 2006). From this point of view, enjoyable, arduous, child-centered math programs and plays can be organized by the means of plays preparing objectives, contents and processes to meet children's needs (Ginsburg et al., 2006).

Math teaching for children is not supposed to be scary. Early math teaching should not be merely focused on making preparations for future problems or difficulties. Teaching children math should bear qualities providing real learning environment that requires challenges and is enjoyable for children and similarly for teachers, and appropriate to development (Ginsburg et al., 2006).

2. Play Centering on Mathematics

Math games can be played with whole class, in small groups and in pairs. Motivating students, games provide high level of attendance and concentration in math class. While supporting learning in different ways, plays can also be used in applying, repeating skills, discovering mathematical relationships and improving problem-solving skill. Parr (1994:29) thinks that the main strength of math games is that they promote solving mental operations, which need to be repeated, and provide opportunities to attain better by practicing whole process over and over. According to another point of view on play, children can experience enjoyable situations where they can control their learning. There is not always only one way to solve the problem and attain success. Despite being designed as individual plays, many math games turn into cooperative plays that children gain great success by supporting each other. These kinds of meticulously planned plays can provide opportunities to improve mathematical thinking skills such as estimation, generalization, proving and explanation (Drews, 2007; Drew et al., 2008). It is known that children really learn daily mathematics by their own attempt. Plays, especially block plays, provide children opportunities open to discovery while forming a basis for more comprehensive activities than mathematical point of view (Ginsburg et al., 2008).

3. Play with the Mathematics That Has Been Taught in School

In early childhood, one of the fundamental approaches may be child-centered teaching. In this approach, teachers should comprehend children's viewpoint, perceive their latest developmental and mental behaviors and promote children to transform math or any subject into teachable form based on those mentioned above. Teachers should perceive what children understand at that very moment and help them how to explain that understanding from a mathematical point of view (NCTM, 2002, p. 6). Plays can be promising arrangements for child-centered teaching (Ginsburg et al., 2006). Stating that traditional math classes that are not enriched via plays, merely taught via worksheets, times table and textbooks cause math anxiety in children. Kamii (1991) discusses that teaching should be supported by daily experiences and math games and also it should be enjoyable in order to make children understand and enjoy math (Taşkın, 2013). Even though math games are beneficial as homework and free activities alone in class, plays planned by math teaching program can be considered to be the most effective ones. Teachers should be precise about the issues such as estimated, aimed learning outcomes of a play, how all students will benefit from that play, what should be done for adults' opinion and support and finally student-child (class) communication (Drews, 2007; Uttal, 2003).

Bruner (1966) mentions three dimensions representing experiences in learning math: (enactive, iconic and symbolic modes). It is accepted to be important in terms of children's developing sense/meaning to express their thought, through using enactive expressions in enactive mode; using painting, emblems, images in iconic mode and language, symbols and signs in symbolic mode. Using physical objects, models and paintings in teaching math can be associated to enactive and iconic modes as well as linking the usage of symbols by the means of mental images and language to symbolic mode. Haylock and Cockburn (2003) emphasize that network connections occurred among perceptible experiences, paintings, languages and symbols are important for interpreting mathematical content (Drews, 2007). The diagram of significant connections in understanding mathematics suggested by Haylock and Cockburn (2003) is given in Figure 3.

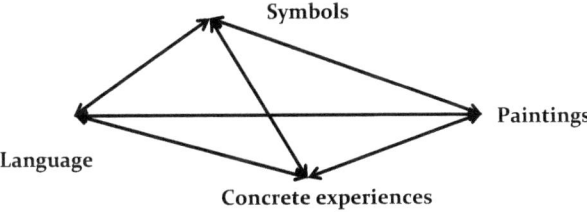

Fig. 3: Significant Connections in Understanding mathematics
Resource: (Haylock and Cockburn, 2003)

In Lesh's Modelling Cycles, a different model in interpreting mathematical content, initial mathematical thoughts are represented in five different ways as seen in Figure 4. They are manipulatives, paintings, real life examples, oral symbols and written symbols. This model developed by Lesh et al. (2003) depends on:

- Children's ability to represent mathematical thoughts via different ways/styles,
- Making connections among different styles,
- The idea that symbolizations to be made among those styles would make thoughts more meaningful.

In addition to Bruner's enactive, iconic and symbolic modes, this approach adds real life examples and oral symbols. In the modelling approach, connections to be made in and among symbols are based on different types of experiences and deep mathematical thinking. Thus, provided transforming and rearticulating a thought flow of thought will be able to present an active learning and teaching environment along with mental development. Just as this model can be acknowledged as a program approach in different classes and math class, where abstraction is deep and profound, it can also be accepted as a teaching program in class or a teaching evaluation model (Lesh, Cramer, Doerr, Post, Zawojewski, 2003). Moreover, this model can guide students to attain mathematical thought and teachers to plan a play and play environment.

Students giving positive reactions to different symbols in planned play environments bring forward some teaching principles. In this sense, it can be said that time of students' transition from concrete manipulatives to semi-concrete manipulatives depend on students' pre-learning experiences and developmental maturity (Samelson, 2009; Morin & Samelson, 2015). The fact that students use manipulatives while solving problem does not guarantee that they will use those

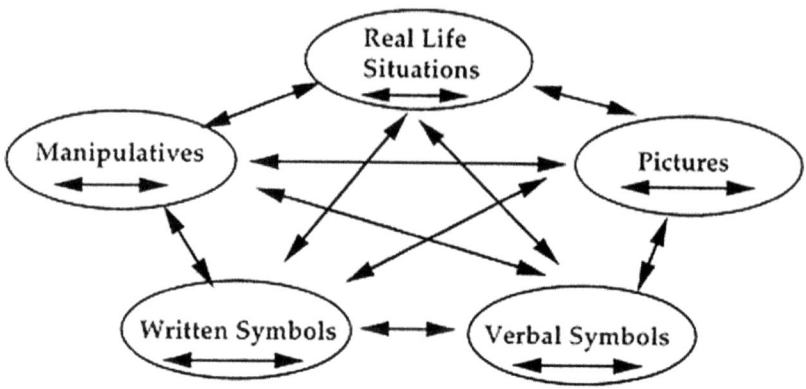

Fig. 4: The Lesh Translation Model
Resource: Lesh et al., 2003.

materials relatively. Even though independent studying (free play) is important for students, first grade students with low oral skills are seemed to be willing to use materials. According to Humbert and Samelson (2010), those students develop various wrong problem-solving strategies while solving questions requiring arithmetical calculations. Those wrong problem-solving strategies can be prevented by means of detailed teaching how to use materials supporting to understand the problem (Morin & Samelson, 2015). While materials incoherent with subject or concept are confusing students, it causes them to feel they could not attain the learning level to reveal their potentials. In this sense, teachers are required to focus on using materials to provide consistence and equivalence with the numbers and operations to be taught and to represent numbers (Morin & Samelson, 2015).

In this context, it can be claimed that Dienes' approach of math and play in terms of manipulatives leads math teaching to a different dimension. Even though Dienes' main area of interest is math teaching in early years, he is considered to be the founding father of using manipulatives and plays in math. His most important approach about this situation lies in the idea that he thinks mathematical concepts as designed and arranged experiences. According to Dienes (1959), experiences have their own objects and some experiences are more inclined to indicate the accuracy of a subject in comparison with other experiences. Dienes thinks that children see and interpret the world differently, and he emphasizes that children's active involvement in understanding the world is significant. He also thinks that students who learn differently should be provided a lot of epitomes/materials to explain the subject rather than a single example. He tries to explain his learning theory developed in this sense by the means of perceptual variability principle (multiple embodiment principle) and mathematical variability principle (Dienes, 1971; Fossa, 2003; Gningue, 2006).

Perceptual variability principle (multiple embodiment principle) is based on the idea that encountering a lot of physical content and materials would help children increase learning. Designing different experiences via many and various materials, students' abstracting math subjects is supported. In this sense, children can use and play with counting sticks, abacus, and ten-blocks or such activities can be organized. Thus, children will be able to make connections between materials and subject and they can abstract mathematically.

For mathematical variability principle, there is the idea that children should be provided with examples irrelevant with the subject or unrepresentative of the subject by means of different experiences to determine the mathematical subject, which is established and steady in all of the manipulation processes related to the subject. Dienes states that the difficulty in learning mathematical subjects lies in the processes of abstraction and generalization. That's why he declares that both of these principles should be used together (Dienes, 1967; Sylvester, 1989; Gningue, 2006).

Dienes states that the first moment that people contact with the real world is important for concreteness, and that people take the first step toward abstraction through interacting with objects and incidents and reacting (Dienes, 1971: 337). Dienes also states that isomorphism is the process of attaining abstractions. Explaining that isomorphism occurs as a result of the fact that some essential forms (morphemes) have come together with others in real world environment, Dienes claims that the main issue in successful complementation of abstraction process is reduction of objects into isomorphic figures. In this sense, **structuralism-based reasoning** is constituted while building morphemes/forms to fulfill our needs and attempts. According to Dienes, abstraction process has structural characteristics (Dienes, 1963).

In explaining our experiences, Dienes mentions two principles, namely principle of dynamism and structural principle. In **the principle of dynamism**, plays structured at first-hand and put into practice are required to include experiences to be acquired about the math subject demanded to be structured. When it comes to **the structural principle**, the structuring process of plays is required to involve primary analysis processes that are not integratedly included in learning of children up to 12-year-old (Dienes, 1960).

Dienes improved three steps related to learning concepts, namely free play, gradual arrangement and final insight (Dienes, 1963), in further years and schematized them as the six steps of learning mathematical concepts (Hirstein, 2008; Dienes, 1967). These steps are summarized below:

1. **Free Play/Interaction**: In a free play, certain arrangement about environment is explored. Play is important to understand and formulate a new subject. In this sense, Dienes mentions two different plays: manipulative and symbolic plays. In manipulative plays, children encounter materials at first hand and try to understand and discover what kind of object a material is. In free plays, stories, different examples and performances are produced about what to do with those materials.
2. **Directed Play/Setting Rules**: Play is turned into a directed play by being structured. Children who get to know materials and interact with them in creative processes in free play act restricted by the features of materials. However, by the guidance of adults, they can continue their play, focusing on the similarities and differences of objects. In directed plays, adults enable students to focus on similar structures/features through the manipulation of objects within the scope of a certain purpose (Dienes, 1967).
3. **Isomorphism**: Plays are compared within the context of same subjects. Subjects/structures are isolated and comprehensive, and isomorphism are tried to be acquired.
4. **Presentations/Representations**: Similar (isomorphic) structures and situations are presented and demonstrated in a way to represent all forms.
5. **Symbolizing**: Represented elements are symbolized and become the object of the study (thing/object is disabled, and study is carried out through symbols).
6. **Formulating**: Presentations are restricted and turned into axioms (Dienes, 1967; Hirstein, 2008; Sriraman & Lesh, 2007).

In this context, Dienes suggests that children should be encouraged to discover materials and practice abstractions during plays, in that way they are required to participate in the processes to enable themselves to have generalizations, use symbols effectively and explain and reflect what they have learnt (Dienes, 1967).

We use plays in teaching math deliberately or not. However, rational and conscious implementations and guidance to be carried out beginning from the initial play environments, toys and play experiences of children may enable children to internalize math from earlier ages and explore the world of symbols. In this sense, diversity and variety of plays applied in childhood, in preschool education and at school life draws attention. If educators and families get to know those plays and use them effectively, it would pave the way for discovering the fun process of understanding mathematical concepts. In sources, it is seen that different plays are used in teaching math.

Play Types in Teaching Mathematics

Plays and activities are special teaching structures where many methods, techniques and mediums related to teaching can be used in different combinations. Within the context of objectives in teaching math, new teaching methods such as active learning, cooperative learning and teaching based on multiple intelligence theory can be used in plays effectively (Uğurel, 2003). It is stated in the resources about different types of plays that building structures like toy blocks (lego), blocks, cubes (Tracy, 1987; Wolfgang et al., 2001; Moyer, 2001; Oostermeijer et al., 2014), jigsaw puzzles (Caldera et al., 1999, Levine et al., 2012), sudoku practices (Baek et al., 2008); sports and physical competences (Hanson and Kraus, 1998; Broh, 2002; Castelli et al., 2007; Thomas et al., 2009; Eveland-Sayers et al., 2009) and music, rhythm and dance (Catterall et al., 1999; Minton, 2003) have a positive influence on students' both academic and math achievement. Moreover, some studies reflected that using technology, computer, Internet, mobile technology as a means of play and fun have a negative influence on academic achievement of students (OECD, 2009; Duman, 2008; Bayraktar and Gün, 2007; Gencer and Koç, 2012; Demir, Kılıç and Ünal, 2010; Gürsakal, 2012; Akyüz, 2013; Usta, 2014). Demir and Yıldızlı (2015) investigated students' game preferences, duration of playing and academic achievement in math. The results of the study showed that there was a significant positive relationship between students' math achievement and playing PC, on the Internet, smart phone, tablet and social media games; however, there was a significant negative relationship between students' math achievement and the duration of playing games. The study also revealed a significant positive relationship between math achievement and duration of playing word games, crossword puzzles, mind games and card game. In other words, the study concluded that the more the duration of playing such games increases, the more academic achievement increases. Furthermore, math achievement indicates similar characteristics in terms of the duration of sports, musical and rhythmic plays, puzzles, blocks, copy-paste games and traditional street games (Demir and Yıldızlı, 2015).

Knowing different types of plays and also their features may be helpful for teachers in developing plays, planning play-based math curriculum, teaching through plays and in applying and evaluating processes. In this sense, manipulative plays, card games and mind games are studied.

1. Manipulative Games

Materials enable children to participate in various processes. Playing materials support both verbal plays like impersonation, imitation and demonstration and non-verbal, manipulative, structural and object plays. It is stated that material use increases the test scores of persistency and problem solving, and physical (concrete) materials used in company with experienced teachers improve students' attitude toward math in a positive way (Sowell, 1989; Clements & Sarama, 2015). However, using material does not guarantee success (Baroody, 1989). In some researches, it is stated that the group that does not use materials is more successful in transfer ability test than the group using materials (Fennema, 1972). Students learn how to use materials by heart, follow the true steps, but complete the activities without perceiving the mathematical thinking that will cause to use materials (Hiebert & Wearne, 1992). Likewise, students may have difficulties in assembling the relation between numbers and blocks in using ten-blocks (Thompson& Thompson, 1990; Thompson, 1992; Thompson & Lambdin, 1994; Clements &Sarama, 2015). Teachers often use materials since examples reflecting mathematical thinking to reform math teaching are few in number and other elements of teaching are required to be changed (Grant et al., 1996).

Briefly, even though researches recommend that teaching should start perceptibly, they warn that materials are not sufficient to guarantee significant/ meaningful learning. In this sense, when manipulative plays are mentioned, concrete materials are brought to agenda. It may be said that concrete materials are involved in teaching and plays via different methods and techniques. In this context, the question what we think of concreteness becomes the main topic.

The concept of concrete means the object students hold. Since this sensorial environment enables one to form a meaningful relation to himself/herself intuitively (feeling the shape, temperature, color, smell and texture), it makes materials "real" in appearance. In his research on Cuisenaire sticks, John Holt (2009) observes a strong relation between the world of numbers and sticks, hence, states that looking at sticks, children can understand how the worlds of numbers and operations function.

Good manipulative materials can provide perceptible information, which makes sense for the learner, grants control and flexibility; is consistent with cognitive and mathematical structure or has characteristics to mirror those and helps the learner to establish a relationship among different information fragments (Clement, 1999). Physical and visual manipulatives create meaningful renditions between the object and action. In this way, students may be enabled to learn mathematical object, thought, action, operation and process. Students need adults' assistance to establish those connections and comprehend those relations (Clements & Sarama, 2015).

While teaching students how to solve a problem, they are expected not only to learn the rules and use them properly but also to understand why rules are useful. The key question on the issue of developing students' mathematical thinking may be that how we can help them in obtaining the rules taught in the framework of concepts. Thus, they can transfer rules in the framework of particular subjects. Unfortunately, children have difficulties in grasping concept learning by the means of procedures (rules/by heart), specially while learning contents with abstract mathematical symbols. When children only confine themselves to memorize operations, this situation causes misunderstandings and errors in transferring rules to different operations (Fyfe, McNeil & Borjas, 2015). It is explained that difficulties in acquiring subjects via mathematical operations stems from mistaught or rareriped abstract symbols. In overcoming this problem, concrete materials (blocks, cubes, object, and models) would be helpful in teaching children math subjects.

It may be said that using concrete materials have many benefits. Concrete materials help to activate intuitive and real-world information during constitute learners' information about abstract concepts, and promote physical activity to improve understanding and having in mind (Fyfe, McNeil & Borjas, 2015). The research showed that building plays like toy blocks and cubes (Tracy, 1987; Wolfgang et al., 2001; Moyer, 2001; Oostermeijer et al., 2014) have a positive influence on students' academic achievements and academic achievements of math. However, using concrete materials does not guarantee success (Baroody, 1989; McNeil & Jarvin, 2007). Many researches claim that concrete materials hinder transfer to new and different situations (for instance, Goldstone & Sakamoto, 2003; Kaminski, Sloutsky &Heckler, 2008; Son, Smith & Goldstone, 2011). Besides, there are some research advocating activities which start with concrete materials and gradually use abstract examples (for instance, Bruner, 1966; Fyfe et al., 2014; Gravemeijer, 2002; Lehrer & Schauble, 2002; Laski et al., 2015). Bruner (1966) claims that new subjects should be taught through processes prepared in three steps (concrete, iconic and symbolic). McNeil and Fyfe (2012) state that through the method called concreteness fading in the study conducted together with associate degree program students, activities organized from concrete to abstract have a positive impact on continuance. Different research emphasize that studies of concreteness fading carried out at primary (Fyfe, McNeil &Borjas, 2015) and secondary schools (Butler et al., 2003) have resulted in positively.

According to the meta-analysis that study 55 research related to teaching with and without manipulatives on the efficiency of materials used in plays, manipulatives only affect teaching under certain circumstances (Carbonneau, Marley & Selig, 2013). In this sense, research mentions four principles in studying efficiency of manipulatives in plays (Laski et.al, 2015).
These are:

1. To use manipulatives constantly in a long period of time, to play.
2. To start with concrete materials representative for the subject, to proceed with more abstract ones.
3. Not using daily materials, which are distractive and prevent to focus on the concept.
4. To define the relation between the concept of math and manipulative clearly.

2. Card Games

Another type of play that attracts children is card games. Card games may be the footballers in the teams that children support, characters in cartoons and movies as well as actors. In these games, children use skills such as matching, score games where the highest one wins, completing team or character group. It may be claimed that player cards or just cards, which many children play on benches at playgrounds, in front of stairs or in the class, hiding from teachers, improve many mathematical skills. Skills and attitudes such as attaining numerical value to the characters, lining them, classifying them based on groups and strengths, keeping the cards they consider to be precious can improve mathematical skills of students. It is stated in the research carried out in this context that students' academic achievements in math vary in respect of the length of playing card games (players cards, animal cards, cartoon and movie characters cards); the highest score belongs to the students who play card games for 1-2 hours, and the success average decreases when they play more than 2 hours (Demir and Yıldızlı, 2015).

3. Mind Games

Confining mathematical skills to only arithmetical skills will have a negative influence on students' perceiving the concepts of time and in the further ages. Improving spatial skills enables children to visualize the space, the concepts of below and above, under and on fade away; and understand the world of computers which is the realm of symbols, signs and patterns. Despite the fact that sorting mind games merely under a category is hard, generally, mind games may be considered to be games played against time and rivals, where mental challenges are profound. We may consider plays like jigsaw puzzles, matching geometrical shapes on cubes, finding the match, completing the pattern or number hunt in the category mentioned above along with the plays such as Sudoku, Jenga and chess. A research found out that the level of secondary school students' academic achievement in math vary based on the duration of playing mind games (chess, sudoku and jenga); the math achievements of those students playing mind games up to 2 hours a day differ compared to other students (Demir and Yıldızlı, 2015). In this sense, it is ascertained in the literature that sudoku games (Baek et al., 2008) and chess (Smith and Cage, 2000; Sadık, 2006) have a positive impact on the skills of solving mathematical problems.

4. Computer Games

It may be said that computer games are now an industry where big budgets are reserved, and is followed by millions of people, the economic value of games almost compete with the budgets of countries. Even though there are many programs used for teaching math in the electronic environment, the most outstanding ones may be thought as the increase in the Quest Schools that mostly focus on programing games and designing new games and the inclination of firms with big budges to creating games and designing games in the framework of educational objections (Minecraft). Although the presence of research abroad is considered to be promising, the situation in Turkey may not be thought as hopeful. While having a computer, technological means and Internet use are stated among the factors related to mathematical success in the literature. The directions of the relationship change based on to what end technological means and mediums are used. Using technological means as a medium of fun (such as chatting, downloading songs, playing games and uploading games) affects mathematical success negatively (Duman, 2008; OECD, 2009; Demir, Kılıç and Ünal, 2010; Gürsakal, 2012; Akyüz, 2013). In other words, as frequency of using technology for fun increases, performance (here achievement) in math decreases (Usta, 2014; Demir and Yıldızlı, 2015).

5. Musical and Rhythmic Plays and Sports

It may be said that activities, dance, rhythmic movements and sportive games are the ones that attract the most attention of children from their early young ages. While using music and math symbols, the language of dance can be physical movements, gestures and facial expressions. Participating in activities such as music, rhythm and dance positively affects the academic achievement of mathematics (Catterall et al., 1999; Minton, 2003). In Turkey, Demir and Yıldızlı (2015) found no significant relationship between such activities and academic achievement in teaching mathematics.

It has been stated in many studies that sports activities have a positive effect on individuals regarding the relationship between students' participation in sports activities and academic achievements (Broh, 2002; Hanson & Kraus, 1998; Castelli et al., 2007; Thomas et al. , 2009; Eveland-Sayers et al., 2009). Although it is stated in the literature that it will contribute positively to the academic achievements of students, no significant relationship was observed between the duration of playing sports games and mathematics achievement (Demir & Yıldızlı, 2015). Similarly, Akyüz (2013) emphasizes that sports activities in Turkey have a negative relationship with mathematics achievement, the reasons for this situation should be investigated and the quality of these activities should be questioned.

The Advantages and Disadvantages of Teaching Mathematics through Play

Teaching math through plays has a structure consisting of development, play and learning principles gathered. Math games have much strength such as affecting physical, social, cognitive and emotional development and using the symbol and language peculiar to math itself. According to Goldstein (2012) and David, Goouch and Powell (2015), plays 1) improve motor skills, 2) sensitize emotions, 3) make thinking emphatically and explaining emotions easier, 4) enable to wait for turns at talking, share and have harmony, 5) provide arranging and sorting, 6) extend the satisfaction from the activities, 7) improve vocabulary, 8) increase concentration, 9) provide mental and physical flexibility, 10) enable to play-act and 11) give chances to express creativity and imagination.

Teaching through plays has weaknesses as well as strengths. The most important one of the weaknesses is not because of plays themselves, but it may stem from the attitudes of families and teachers against plays, who have a negative influence on teaching through plays. In the literature, teachers' opinions about play vary based on cultures. Izumi-Taylor et al. (2004) examines American and Japanese teachers' understanding of play and concluded that teachers in both countries think plays are necessary, being coherent with the current spirit of education developmentally, and their understanding of play is related to the cultural structures of their countries. Japanese teachers prefer plays serving as a preliminary preparation for the needs of the group in class, and define plays by the aspect of emotions, passions and attitudes representing the soul, the spirit of life. And it is observed that American teachers relate plays to learning and development, make less time for free plays compared to Japanese teachers. While Western teachers center upon the educational outcomes of plays, Asian teachers focus on plays in spiritual integrity, providing happy, comfortable and enjoyable environment, namely on free plays more (Izumi-Taylor et al, 2004; Izumi-Taylor et al., 2010). In another approach, it is pointed out that teachers' understanding of plays, especially manipulative ones affect the educational contribution provided to students to be either positive or negative, and teachers feel themselves inadequate to use those materials (Tran, 2015; Pham, 2015).

Researches about the efficiency of plays on teaching and different fields (disciplines) gradually increase. However, besides the strengths of plays, misusing or applying them in a lacking way, implications for fun without having enough experience can be thought as the weakness of this technique. There are some warnings in the research carried out like math games does not guarantee teaching (Ginsburg et al., 2006), math games are necessary but not efficient (Morin & Samelson, 2015) and using materials in plays does not guarantee learning concepts (Baroody, 1989; McNeil & Jarvin, 2007). It is seen that students may mislearn or develop misconceptions owing to the use of materials not representing concepts. The fact that using play effectively requires a long time and effort may be thought as the weakness of plays. Besides, while planning games, involving activities higher or lower than the level/grade of students may prevent them from continuing the play. This situation may cause the behaviors that would be acquired through plays to be ignored.

One of the weaknesses of teaching through plays may be that activities last long and the concept of class disappears. In a normal school program, the bell ringing, the noise outside or those involved in the class may reduce or decrease the performance of playing activity. Even though playing activities are applied in long period, their impact on continuance in terms of teaching and achievements in sensory field may ensure to ignore that weakness.

Plays may be considered to be weak to meet the needs of teachers with regards to planning, arranging material and environment, implementation, evaluation, time, expenses and plentitude of teaching objectives. However, this situation will change once teachers experience teaching through plays and as a result of it, evaluate developmental outcomes. Children's changing attitude against classes, school and friends and positive behavioral changes will draw attention as the most important strength of plays.

Teacher's Responsibilities in Teaching through Plays

Curriculum developers that center upon qualified playing experiences of students on streets and at homes mention principles supporting pedagogy and processes of program development and evaluation. These principles may be listed as giving children enough time in a play; creating problems and trying to solve them; providing adult and children participation and arrangement; organizing creative and imaginative activities to support development and learning. Besides that, Wood (2008) mentions following principles on the roles of educators in plays:

1. Planning and creating the learning environment that requires challenge.
2. Supporting children's learning by free and planned plays.
3. Enriching and improving children's vocabulary through the communication in plays.
4. Being sure about the play continues and advances.
5. Observing and evaluating children's plays.

1. Planning and Creating the Learning Environment that Requires Challenge

Teachers may help students learn by arranging the learning environment and planning play processes. Learning environments should support children's socio-emotional, cognitive and physical development. Moreover, learning environments should present cognitive duties that will enable them to discover and have experiences and also use motor skills important for their future lives (Shipley, 2007). While planning play centers, teachers need to take the following requirements into consideration (Demir, 2008; David, Goouch &Powell, 2015).

1. They need to decide on developmental objectives and behaviors.
2. They need to know learning principles and students' learning styles.
3. They need to design and create learning centers.
4. They need to evaluate learning environment.

Qualified plays are associated with cognitive, emotional, social and psycho-motor learning outcomes and positive outcomes in the subfield of learning. In England, Wood (2008) mentions teaching principles that are still being applied related to the combination of teacher-oriented/child-oriented activities and structured/free plays. In this context, the indicator of an effective pedagogy is considered as:

- Giving an opportunity to **the sustainable thinking share** structured by adult and child together,
- **Activities** enabling the interaction of adult-child, and
- **Free play** activities.

In this sense, the role of educators can be founded as creating proactive play/learning fields and meeting children's preferences, interests and learning models. These educational suggestions are actualized by means of sociocultural learning theories (Wood, 2008). While students' interests are centered in teaching plan, subjects in the field of discipline expand and enrich students' learning. Although today's program models handle plays in the context of integrated teaching approaches, it is known that especially in England, qualified plays gain remarkable success in an environment where teachers encounter with responsibilities, performance and success demands, and nations compete about effective learning and teaching. There are considerable evidences about the fact that children improve verbal communication, high level social interaction skills, using game materials creatively, improve imaginary and divergent thinking and problem-solving skills (Wood & Attfield, 2005).

2. Enriching and Improving Children's Vocabulary through Communication in Plays

When adults interact with children during a play, they can comprehend children's interest, information and language developed about plays much better. Educators have some roles and responsibilities before and during a play where interactions occur. These are:

Educators' responsibilities during a play may be listed as follows: to appreciate the play; to be sure about children's safety; to observe the learning; to save and interpret; to evaluate; to establish and start meaningful communication and to create positive classroom atmosphere. **Educators' roles during a play** may be defined as follows: arranging materials, resources, emotional climate and time by organizing the physical environment; being a negotiator in conflicts and disputes and structuring and supporting the learning; expanding and supporting the game. If we summarize the rules, teachers apply in plays:

In sum, **the rules that teachers are required to apply in plays** are listed below:

1. Children should decide the activities themselves.
2. Environment should be equipped with game types that will be meaningful and accessible to children.

3. By observing children, they should plan experiences in a play superior to or above their actual developmental period.
4. They should plan various experiences in a play from simple to complex, concrete learning environment should be arranged to observe concepts and skills.
5. Structured and open-ended activities should be presented in a balanced way.
6. Individual and group activities should be applied equally.
7. They should place equipment, materials and resources properly in learning centers, and coherence should be provided between the center and materials (Shipley, 2007; David, Goouch & Powell, 2015).
8. Teachers should pay attention to their own and children's points of view.
9. Students and teachers should participate to the playing process together.
10. Teacher's guidance and sensitivity related to objectives should enable children to play by themselves.
11. Student-teacher and student-student interaction, communication, principles of creativity, power-opportunities-freedom choice should be taken into consideration (Samuelsson & Carlsson, 2008).

3. Supporting Children's Learning by Free and Structured Plays

Gordon and Browne (2004; 419–420, cited in Imenda, 2012) mention that teachers have two vital roles free plays as 1) simplifying and directing a play and 2) creating the parts and atmosphere of a play. In this sense, experienced teachers:

- Incline children to play and involve while guiding a play, instead of leading the game directly or taking on tasks,
- Benefit from students' opinions and thoughts,
- Do not impose any kind of thought upon children,
- Show the behaviors of special characters such as how to use an equipment and how to be the next in a line in plays when necessary,
- Raise questions in need of interpretive answers about what is going on,
- Help children start, finish and continue the play,
- Give hints to continue the play,
- Enable children to focus on each other's behaviors, and support their interaction,
- Announce children's behaviors loudly to the class when necessary,
- Help children understand each other's feelings and vocalize them while overcoming conflicts or discussions,
- Provide the play to be improved by asking questions to support discovery and inventions (Imenda, 2012).

4. Observing and Evaluating Children's Plays

There is an active learning process ongoing in the period of observing and evaluating children's plays. While testing the consequences of their behaviors, students experience different plays. Meanwhile, the duty of teachers will be to guide students in the framework of active learning process and prepare the next step as a result of the evaluation. In this sense, active participation in the play for all students is in question. To reach this participation to the desired level, principles that will make adult' observation and evaluation processes easier can be determined.

To enable all students to play, adults can be supportive by:

- Changing social and physical environment,
- Reflecting quantitative and qualitative experiences to fulfill more than one sense, changing and increasing the activities appropriate for children,
- Using cooperative learning in accordance with the potentials of all children,
- Separating activities into small parts and restricting the instructions given at once,
- Using the activity, friend or material that children prefer,
- Participating plays, practicing example and model studies more than one, giving motivational support,
- Using photographs and jigsaw puzzles with big pieces for children with underdeveloped verbal skills.

Teachers' responsibilities may be planned not only by means of teaching method but also the discipline taught. In this sense, there are some principles for teachers to be taken into consideration in the context of teaching math through plays. According to Anthony & Walshaw (2009), effective **math teaching:**

1. Accepts to improve positive mathematical identities of all students, regardless of their age, and turn them into effective math learners.
2. Is based on interpersonal respect, sensitivity; on being sensitive about various cultures, thinking processes and realities of class.
3. Focuses on restoring a range of academic outcomes such as conceptual understanding (perceiving), operational fluency (implementation), strategical competence and flexible reasoning (high-level skills).
4. Is determined about increasing a wide-range social communication, which will make a great contribution to the integrated development to make students productive citizens in math classes (Anthony & Walshaw, 2009).

In this sense, effective math teaching is tried to be explained via the principles namely highlighting the ethical values in classes; preparing the learning environment; mathematical communication; mathematical language; beneficial tasks; making connections and teachers' knowledge.

1. **Highlighting ethical values in class:** Groups in a class should be given importance that will help students improve their mathematical proficiency and identity, and focus on mathematical objectives (Anthony & Walshaw, 2009).
2. **Preparing the learning environment:** Influential teachers provide opportunities to make sense of students' individual and cooperative thoughts. Another important duty of teachers is that they provide arrangements to fulfill students' needs. Some students want to study individually in silence while others prefer cooperative environments where ideas are shared and discussed. Teachers must pay attention to those needs in class. Influential teachers begin structuring students' thoughts from their current competence, interests and experiences. As students become more experienced, methods such as creating barriers before solving a problem for students, omitting some information, making them use symbols and representation, and generalizing may be used (Anthony & Walshaw, 2009).
3. **Mathematical communication:** Influential teachers make classroom communication based on mathematical proof easier. Students should be trained on how to explain mathematical definitions and operations and prove the results of problem they get. In this regard, verbal, written and concrete presentations should be made and activity process must be continued with explanation proving and guidance processes. Teachers may provide classroom interaction by repeating what is uttered, paraphrasing it or expanding student discussions. Teacher may use the following re-voicing techniques:
   - Repeating proposal or thought of student,
   - Explaining the hidden meaning in a student's thought,
   - Discussing the definition or meaning with a student,
   - Bringing forward new ideas or opposite opinion (Forman & Ansell, 2001).
4. **Mathematical language:** Mathematical language is formed when it is modeled once teachers provide proper conditions and enable students to engage in dialogues they can understand. Math teachers use large-scale formal and informal evaluation techniques to observe the process of learning, to test learning and to determine what can be done and improved. In daily activities in classes, teachers gather information on how students learn, what they know and can do and what their interests are. In shaping this data, mathematical language that students use in group or individual studies is instructive for teachers. Feedbacks given by teachers are important in creating

mathematical language. On what is wrong and right in their classes, influential teachers on teaching, progress via feedbacks and corrections going on step by step. Influential teachers go on teaching step by step using feedback and correction. Teachers provide chances to improve mathematical language making students write exam questions, success criteria and math articles to evaluate their own studies (Anthony & Walshaw, 2009).

5. **Beneficial tasks:** Influential teachers have understood how chosen tasks and examples affect the way students consider, improve and use math. Tasks applied to open-ended model are not only used in applying math but also enable many problem-solving strategies and new mathematical thinking to be emerged. Through questions related to daily life, students get to know social life as well as notice the importance of math for social life and other disciplines. Thus, mathematical problems are supposed to be arranged in a way to visualize problem via a model and solve it by doing and living it, rather than tending to give the right answer (Anthony & Walshaw, 2009).

6. **Making connections:** Influential teachers make connections among different problem-solving techniques of students, math subjects and daily mathematics experiences. Influential teachers use many means and representatives to support students' mathematical thoughts. These means may be number systems, symbols, graphics, diagrams, models, equations, formulations, metaphors, narrations, textbooks and technologies in order to make sense of and expand mathematical reasoning (Anthony & Walshaw, 2009).

7. **Mathematical equipment and models**: Influential teachers decide on means and models carefully to support students' thoughts. Teachers using equipment/means effectively to support mathematical reasoning and interpretation has an important role as well (Blanton &Kaput, 2005). Using multiple symbols and models will help students improve their conceptual and operational flexibilities. Means are quite helpful by the aspect of discussing thoughts which are hard to explain and write. Receiving support from narrations, drawings, symbols, concrete materials, visual manipulatives on this matter, teachers use them to explain their opinions (Anthony & Walshaw, 2009).

8. **Teacher's knowledge:** Influential teachers improve and use thinking out loud which will enable learning, and take all students' mathematical needs into consideration. How teachers organize their classes, how much teachers understand math, what they believe about math are all related to what they understand about math teaching and learning. No matter how good their

intentions are, teachers should work in a way to enable students to acquire fundamental mathematical thought. In order to achieve that, their studies should depend on pedagogical information and implementation related to the field (Anthony & Walshaw, 2009).

In addition to teaching principles, teachers need a well-equipped environment and effective implementations to support teaching through plays. Implementations to be carried out in this context may be stated as follows:

1. **Observing children's plays**
   If children's toy blocks plays are disconnected and discontinued, you may share books and illustrations with block models.
2. **Being gentle while intervening**
   One of the most beneficial techniques is to decide on whether social interaction and mathematical thinking is improved or not. If there is a progress, just make observation and leave children alone with their studies/practices. Allow them to share their experiences later in all class activities.
3. **Discussing opinions, clarifying and explaining**
   While children are having a discussion in block games about which one is taller, when you see one of them mention height and the other one mentions width, explain them how to consider buildings big in various aspects. You may have explanations such as one is taller than the other, and the other one is taller.
4. **Planning long period of times for plays**
   Teachers should provide rich learning environments to improve mathematical thinking, equipped with structured materials like toy blocks.
5. **Encouraging ongoing activities and plays**

   Children develop mathematical discoveries thanks to challenge, guidance, and proper task and communication skills provided by their experienced teachers (Clements & Sarama, 2005).

Research on Teaching Math through Plays in Turkey

When research in Turkey are taken into consideration, it is seen that research about plays started at the end of 90s, gained interest gradually, and conducted by means of candidate teachers, teachers, parents and different age groups. In a research, teachers indicated that they have not had adequate classes about plays at the university they graduated from, they could not reach the sources and publications to educate, improve themselves. There is a standpoint shared that teachers do not spare enough time for teaching through plays and do not use teaching through plays as a method of teaching while planning the class, and plays are slightly involved in elementary curriculum (Demirci, 2004). It is pointed out that teachers abstain from preferring plays because they have inefficient information about teaching through plays, intense curriculum, examination system and physical and social structure of school. When connotations of play and math are taken into consideration, teacher candidates mostly respond with the word "harmony" while teachers have chosen the phrase "being interesting" (Uğurel, 2003). From the standpoint of this situation, it may be claimed that teacher candidates are more ready to learn through plays; however, teachers remain distant to plays in math class. In Gunes's research (2010), it is concluded that teachers approve to use plays and activities in math classes for secondary school education, but they face with various problems; finally, teachers at a consensus that school potentials and curriculum must be adapted to the use of plays and activities, and this method would be more productive through practices such as joining in-service-trainings. In this sense, it may be claimed that teachers are required to be trained about teaching through educational plays; publications and books about theory and implementation of the method should be prepared; sample classes should be held; and teaching programs to carry out those implementations for each age group should be included.

When it comes to the experimental research conducted by means of teaching method through plays in math class, it is seen that different variables are studied, namely achievement, retention, attitude, problem-solving skill, success, self-regulatory learning strategies, motivational beliefs and self-efficacy perception. When the impact of teaching math through plays on students' achievement in math is taken into consideration, it is concluded that teaching math through plays for first graders provide a higher math achievement compared to the traditional methods, and rewards take a positive role in teaching math through plays (Kılıç, 2007). Kılıç (2010) and Boz (2014) observed that there was an increase in the achievement level of students in the experimental group applied teaching through plays compared to the control-group taught by traditional approach. Yiğit (2007) examined teaching math through plays for second graders, but there was not a significant difference between the experimental and control group. Tural (2005) found a significant difference in favor of experimental group in which teaching through plays was applied on third graders' math achievement. Altunay (2004) and Yumusak (2014) revealed a significant difference in favor of experimental group in which teaching through plays was applied on fourth graders' math achievement. Biriktir (2008)'s and Yılmaz (2014)'s research with fifth graders, teaching through play made a significant difference in achievement. Aksoy (2010), Gökçen (2009) and Konak

(2009) concluded that teaching through plays had a positive impact on academic achievement sixth graders' math achievement. In addition, web based-math games were more effective on academic achievement (Tural, 2012) research. Canbay's (2012) research with seventh graders concluded that there was a significant difference in favor of the experimental group in which teaching through plays was applied. Songur (2006) with eighth graders revealed that there was a significant difference for the math achievement of experimental group in which teaching through plays and puzzles was applied, the method increased the perceived success levels in math and exerted an influence over the perceived benefits of math. Ultimately, based on research, it may be said that when academic achievement and retention scores are taken into consideration, teaching math through educational plays is an effective method.

When retention is taken into account, Songur (2006) indicated that teaching math through plays made it easier for with eighth graders to remember what they learnt in math classes; likewise, Gökçen (2009) observed that teaching through plays had a positive impact on retention of learning outcomes acquired in sixth graders' math classes. On the other hand, Konak (2009) concluded that there was no significant difference in terms of retention in sixth graders' math classes. Yumuşak (2014) found out that teaching through plays provided retention in fourth graders' math classes. Nevertheless, Yiğit (2007) revealed that there was no significant difference between experimental control and group in terms of retention in second graders' math classes. When the research are considered, it may be claimed that teaching through educational plays provides retention. However, it may be said that this conclusion needs to be studied more in different classes and via further research.

Interest in play-based math teaching in Turkey is increasing every year. Play-based articles are remarkable in different disciplines and specifically in the field of mathematics education. Qualitative and quantitative developments in play-based content, online game sites, videos and books are pleasing. At the same time, it can be said that the increase in the types of games played at home like board games with Covid-19 pandemics allows children to meet in these games from preschool and offer opportunities to a rich learning environment.

Research Abroad on Teaching through Plays

While research abroad is examined, it is seen that teaching math through plays increasingly draws attention of researches. Learning through plays was paid attention in 1850s in resources abroad drew more attention of teachers and researches since 1900s. When it comes to 2000s, developing technology and restricted city life increased the requirement of teaching methods to fulfill social, psychological, physical and spiritual needs of students. It was observed that children used math beginning from early ages, but only as a result of their communication with their elderly, they could improve it. In this sense, math can be introduced to children through plays and caused to be loved by children and children are helped to overcome problems in their daily lives. Based on this idea, it is seen that interest of research focuses on subjects like math teaching in early ages, play and math, impact of mathematical achievement in early ages on further social and economic success. While structural and social constructivist approaches have been maintaining to be influential, it may be said that the idea claiming students in preschool period and earlier ages, who are not in concrete operational stages, can learn math to a certain extend attracts supporters gradually. It is seen that along with the symbolic plays advocated by social constructivists and especially the use of manipulatives and visual manipulatives, plays aimed at developing spatial, communicative, problem-solving, estimating and decision-making skills of students have been designed. It is seen that research abroad are conducted through different variables instead of academic success on the contrary to the research in Turkey, and qualitative research gain importance rather than quantitative ones.

Cheng and Johnson (2010) examine four education and four early childhood journals between 2002 and 2007, and they encountered the word play in the titles, abstract and key words of only 57 articles out of 1000. While 16 of them primarily focus on play, only 7 are written on education through plays. In this context, play is studied under four different roles as a) a subject/content, b) having a great role, c) having a minor role and d) being related to intervention and special students (Cheng & Johnson, 2010).

Sedighian and Sedighian (1996) point out that play maintains meaningful learning; it has objectives and grants success when objectives are realized; it requires challenge with difficulties; it promotes creating cognitive products; it is linked with mathematical concepts and it motivates and functions as a sensory stimulus while enjoying oneself.

Raymond (1997) studies relationship between beliefs and implementations of first grade teachers for 10 months. The study uses interviews, observations, data analysis and survey of beliefs as a data collection tool. It indicates that implementations of teachers are mostly related to their thoughts on math content rather than thoughts on math teaching, and this situation is connected with math experience of when they were students (cited in Taşkın, 2013).

Wolfgang, Stanard and Jones (2001) conducted a longitudinal research on 37 children at the age of 4 to study the impact of block games on mathematical success. While children were enrolled in kindergarten at the age of 4, first some tests were applied. The same children were monitored until they graduated from high school. Research findings indicated that playing with blocks in preschool increased mathematical success in future education life. Furthermore, a strong correlation between playing with blocks and mathematical success at primary school was not found. A strong link between block games and math achievements like math-weighted classes beginning from the seventh grade at secondary school and during the whole high school period, number of rewards, high marks they got in math class was discovered. A link among the scores they got from standard tests was detected as well.

In his Ph.D study, Badzis (2003) discusses the impact and role of play on children's learning in preschool education from the perspective of teachers and parents. To this end, teachers' viewpoint of play is studied by means of their play definitions, their thoughts on the value and role of plays in learning and their practice of play in teaching. Parents' perspective of plays is investigated whether they accept plays as a pedagogical tool or not and their preferences of plays. Data gathered through semi-structured face-to-face interviews with 30 teachers, 30 parents, 15 administrators, 12 students and an curriculum developer are reinforced by observations made in 15 preschool institutions of 5 different kinds. According to the findings of the study:

1. Plays in education and child development do not compromise with teacher perceptions.
2. Few parents consider plays essential in learning. Many parents prefer formal education environment.
3. While plays are not considered to be able to provide learning experience in many implementations, most of the teachers teach following a formal education.
4. While there is no significant difference between implementation and philosophy of teachers related to their perspective of plays, it is recognized that there is difference in teachers' knowledge level, working hours and academic levels.
5. In preschool education implementations in Malaysia, 4 fundamental factors are mentioned, which prevents improvement and use of plays for teaching. Those may be stated as contextual obstacles related to plays, obstacles related to attitude, structural obstacles and obstacles in implementation. To conclude, it is explained that implementations need to be reviewed and

approaches should be developed in order to both apply plays as a method in learning and to deal with stress (Badzis, 2003).

Christakis, Zimmerman and Garrison (2007) investigate the relationship between language development and children's playing with blocks was discussed. Randomly selected experimental and control groups consisting of babies at the age of 1,5–2,5, of a low and middle-income families were created. In the beginning of the study, the babies in the experimental group were given two sets of blocks. Parents of those babies were told to encourage their babies to play with blocks. Control group babies were not given any blocks and their parents were not told to do anything, either. Parents in both groups were asked to keep a journal about the activities of their children. The real objective of the study was not revealed to parents, but they were only told that they were the part of a study on how their babies spend their time. Six weeks later, an interview with each parent, including a language development evaluation of their children was done. To conclude, it was indicated that babies/children in block group got higher marks in verbal comprehension, grammar and vocabulary tests compared to the babies/children in the control group, and even though it was not significant, the ones in the experimental group started watching TV less. Although a certain reason was not indicated about why block games had such an influence, it is possible to have such a consequence because babies playing with blocks interacted and spoke with their parents more. Or, playing with blocks itself seriously reinforces children's language development.

Math games are commonly used as a reward for students who finish activities early or for increasing student's attitude towards math. In the study of Bragg (2007), 222 fifth and sixth grade students were taught multiplication and division of decimal numbers by means of calculator plays and enriched math activities for 4 weeks. At the end of the study, it was stated that students adopted negative attitude in an unexpected level about considering plays as a tool to learn math. While contradictory methodological dilemma occurred against play was being explained in this study, concerns about this dilemma containing controversy was explained as follows:

1. By the aspect of students' attitude against play, it was stated that children did not consider play as a learning tool. Past experiences and presenting play as a rewarding activity done when the class was about to end were considered to be reason of the conclusion above. In this sense, plays were recommended to be organized as activities of practice and discovery in depth.
2. Another finding in the research showed that scales on attitudes might not have been understood by students. It was stated that students had hard times

in understanding the Statement "math games help to learn math." Moreover, it was mentioned that some words in attitude scale such as metacognitive, projection and self-awareness were abstract.
3. The implementation order of the scale might have caused this negative attitude. Carrying out as a first implementation before the achievement test might have caused students to give negative responses to the attitude test.

As a conclusion, how to measure and explain the data by benefitting from various perspectives is stated in the research. What the research lacks may be defined with the situation that students do not have terminological competence related to the attitude scale and past experiences to consider play as a teaching tool (Bragg, 2007).

In the research on whether symbolic plays prepared through a fictional situation help to increase mathematical activities of preschool students, Edo et al. (2009) studied 13 different symbolic plays such as designing a store, designing a pastry, naming the pastry, visiting the pastry, trade and shopping, using a calculator, which were planned on the play corner in the class for 26 students at the age of 5-6 at a state school near Barcelona. The main objective of the research in the framework of those plays was:

- To reinforce observation, analysis and learning skills students use in daily life.
- To organize cooperation and coordination skills within the processes of putting forward an idea, sharing duties, reaching a consensus and then working for a cause.
- To use cultural tools at school and in daily life, namely reading, writing, speaking, illustrating and modelling.

Data was gathered in the research through studying class observations, video and audio records, interviews and students' pictograms. While interviewing 26 students before and after the play, products of each student were analyzed one by one. As a result of the analysis of class observations and student products conducted in the context of 13 different plays, findings within the scope of five different themes were acquired. During the plays, students went through phases like:

1. Using nonnumeric quantitative expressions (few, little, many, more),
2. Using quantitative meanings of numbers unconsciously (telling the number but without thinking its quantity),
3. Using quantitative meanings of numbers and nonnumeric phrases (without an operation),

4. Using arithmetical operations with numbers (addition, subtracting), and
5. Using a calculator.

What is learnt at the end of the research is as follows:

1. Making a choice, counting votes and expressing ideas and results democratically,
2. Observing and analyzing how to use money,
3. Reading numbers and putting forward ideas about their quantity,
4. Thinking about the value of objects (products), reaching a consensus and using the skill formerly acquired,
5. Using a calculator and developing approaches for addition,
6. Reading and writing many numbers within an end and a context, and
7. Learning and using math in an enjoyable way under the guidance of play with a rich content (Edo et al., 2009).

Sarama & Clements (2009) studied how plays can improve children's scientific fundaments and basis in learning math and how adults can support students' performances and mathematizing. At the end of the research, it was observed that children improved mathematical thinking and causation skills, especially when they had enough information about plays and materials, when rules and process, namely the extent was clear and incentive, and also when the content was familiar and safe for the world of children, they could improve those skills better. It was pointed out that math could be smoothly integrated with ongoing plays and activities of children; however, it was required for adults to have enough knowledge, supportive content and language. It was stated that these mediums would be reinforced toy blocks, building toys, card and puzzle games, computer games and other materials. Children benefitted from rich playing experiences by being prepared for future math learning and exploring new ways to interpret their own world (Sarama & Clements, 2009).

In a research by Lee & Ginsburg (2009), it is indicated that teachers have nine misconceptions related to math. These are as follows:

1. Little kids are not ready for math education.
2. Math is only meant for gifted children with math genes.
3. Simple shapes and numbers are enough for math teaching.
4. Language development and reading-writing is more important than math.
5. Teachers must provide an enriched physical environment, remain behind, observe children and let them play.
6. Math must not be taught as a subject separately.

7. Children can learn math through interacting only with concrete materials.
8. Mathematical evaluation is not important when it comes to the little children.
9. Computer is not appropriate for math learning and teaching.

Related to the consequences of a long-termed research about the play-based program approach, Van Oers (2010) observed 5- to 7-year-old children. In this sense, 34 students played games related to numbers within the context of the play-based program approach. Findings about mathematical achievements (counting skills, standard test achievement) were above the national success average of the children at given age range. Furthermore, the children in the experimental group were not applied any teaching related to those subjects. In this context, it is stated that play-based program may be considered to provide a good medium related to the concept of number and counting at an early age. Another finding is that program enables a chartered teacher to develop playing activities and use different applications, being flexible to use the program for higher grades. Activities of counting skills may be transformed into four operation activities via different implementations in the play (Van Oers, 2010).

In Chen's (2011) doctorate study, teachers' perspectives and implementations related to learning through plays in preschools where 4–6-year-old children are taught in Singapore are studied within the qualitative pattern. Chen points out that plays are considered to be categorically enjoyable based on focus group discussions and findings acquired from data gathered via literature reviews. It is seen that the themes invented about what plays are: "play is entertainment," "play is freedom of choice" and "play is volunteering." Moreover, it is stated that play provides learning opportunities by practicing, through peer interactions, granting child-centered, process-oriented and guided discoveries. In another sub-problem, by gathering teachers' responses related to the benefits of play under certain themes, contributions of play to cognitive (academic disciplines and problem solving), social (waiting for one's turn and studying/acting together), emotional (self-confidence and having confidence in others) and physical development (gross and fine motor skills) are presented. In another sub-problem related to teachers' roles, it is found that teachers take the roles of participating plays with children (guiding behaviors, being a role-model, mutual respect, being playmates and being facilitator) and providing coordination with families as a reflective and critical pedagogue (taking the role of observant, planning and evaluating) and educator. At the final sub-problem of the research, teachers' standpoints related to using play as a learning tool are discussed. Three standpoints defined concerning this sub-problem are: restrictions (time, budget, sources, playing field and education of employee), class management (children's behaviors and composition and arrangement of the class) and attitude confinements (family expectations, peer support, administrative supports and expectations of administrators and administration) (Chen, 2011).

In doctoral study conducted by DeGroot (2012), the program of "math play" was constituted on how to support, report and evaluate development of math in play-based preschool education environment arranged by Reggio Emilia principles. As well as including an environment to support cognitive development and activities appropriate with children's development, this program explains how to report initial educational themes in class in a way not to be contrary to the principles and implementations of the approach mentioned. Considering mathematical development, math Play Program successfully reported how every child understood math and which fields need to be improved. Moreover, in the reporting process, teachers were provided examples on how to evaluate via portfolio templates. Math Play program also provides an alternative evaluation scale to the standard tests where first-hand evaluations can be done. Based on the findings following the implementation of this program:

1. Participating in activities, children made a mathematical progress in different fields fulfilling 18 sub standards of the roof program of California preschool learning society.
2. It was observed that portfolios prepared at first hand were effective methods of deciding on individual mathematical progress of children by the help of teachers and parents.

Furthermore, it was observed that mathematical skills of students participated in the program along with the activities applied continued even after the doctoral study was completed (DeGroot, 2012).

In the opinions of teacher candidates related to the practices on enrichment of math teaching through plays intended for teacher candidates, Meletiou-Mavrotheris and Mavrotheris (2012) state that those practices can be considered as enjoyable tools to be used in math education, make classes free from dullness and turn them into classes children do not get bored, draw students' attention in learning mathematical concepts and help students understand those concepts.

Chapter 3 Method

Abstract In this chapter, research design, study group, data collection tools, data analysis, course of research and games applied are elaborated. The final part includes findings related to hypotheses and sub-problems.

Research Design

Explanatory research approaches and mixed method (quantitative and qualitative) were used in the study. Pre-test–post-test matched control group design was used as part of the quantitative approach during the study. The experimental process lasted 8 weeks. Pre-test and post-test measuring addition and subtraction objectives in scope of the unit "Whole Numbers" were applied in both the experiment and control group. Three weeks after the post-test, the same achievement tests were applied in the groups to determine the retention of learning.

In the qualitative research of the study, one of the qualitative research designs, case study design was used. Case study; It is an empirical research method that works on a current phenomenon within its own life frame, and is used when the boundaries between the phenomenon and its content are not clear and where there are more than one evidence or data source (Yıldırım & Şimşek, 2005: 277). Descriptive analysis techniques were used in the analysis of qualitative data obtained from the student interviews, classroom observations and open-ended inquiry questions during and after the application process.

Study Group

The study group consisted of 54 first grade students in a public primary school, in Gaziosmanpaşa, İstanbul in the 2014–2015 academic years. All first classes and classroom teachers were randomly designated by school administration. Based on this random selection experimental group included 27 students (12 female, 15 male) and control group included 27 students (13 female, 14 male). The school consists of classes of 30–35 students in an environment where more than 1200 students receive education, where workers, tradesmen and low-income families of middle and lower socio-economic levels migrated to the city densely from the Black Sea, Central Anatolia and Eastern regions. The

Method

Tab. 3: Diagram of Quasi-Experimental Design of the Study

Groups	selection	Pre test	Method	Post test	Retention
G1(experimental)	random	Q1	x	Q1.2	Q1.3
G2 (control)	random	Q2		Q2.1	Q2.3

Tab. 4: Age Distribution of Study Groups

Age-month	N	X	Ss	Sd	t	p
Control	27	77,30	3,97	52	-0,538	0,593
Experimental	27	77,89	4,11			

Tab. 6: Reading Comprehension Test Scores of Study Groups

Reading Comprehension	N	X	Ss	Sd	t	p
Control	27	90,24	13,20	52	-0,13	,896
Experimental	27	90,81	18,50			

Tab. 5: Mathematics Pre-Test Scores of Study Groups

Math pre-test	N	X	Ss	Sd	t	p
Control	27	12,48	7,54	52	-0,83	0,409
Experimental	27	14,30	8,44			

evaluation of the study groups in terms of age, reading comprehension and mathematics pre-test scores is given in the Tables 4–6.

As shown in Table 4, (t (52) = -0.538 p> 0.05) by looking at the p> 0.05 and t value, it was seen that there is no significant difference between the groups of means of the study groups' age.

As shown in Table 5, (t (52) = -0.83 p> 0.05) by looking at the p> 0.05 and t value, it was seen that there is no significant difference between the math pre-test results of the study groups.

As shown in Table 6, (t (52) = -0.13 p> 0.05) by looking at the p> 0.05 and t value, it was seen that there is no significant difference between the comprehension test scores of study groups.

Data Collection Tools

In the quantitative aspect of the research, an achievement test developed by the researcher. The achievement test included questions at the level of knowledge, comprehension and application in the unit titled "Numbers." The achievement test included 35 open-ended and multiple-choice questions. The achievement test was used as pre-test, post-test and retention-test. For the validity of the test, opinions of 4 math education experts, two education experts and three teachers were taken, and the correspondence value among those experts was calculated as 0, 93 (Miles and Hubermann, 1994). Carrying out the pilot implementation of the test, reliability co-efficient was calculated as r=0,88 based on KR-20 method. Values of the item difficulty index of the test (pj) was observed to be between 0,11 and 0,61. The total of difficulty index of items present in the test was divided by the numbers of items; thus average difficulty of the test was calculated as 0,36. It may be said that medium and hard questions considering difficulty are equally distributed at the levels of information, comprehension and implementation in the test. Item discrimination indicates to what extent items can discriminate individuals related to the quality measured. It is the competency to discriminate between the individuals having the quality aimed to be measured by the test to a large extent and individuals having that quality in small measure. Item discrimination co-efficient or index may have values between −1.0 and +1.0. Items with a value of 0,4 and higher were defined as very good items, whereas values of 0,3 and below, negative values to be corrected, were defined as items to be removed from the test. When item discrimination of the test was calculated, no item with negative value was found, whereas, it was seen that, nine items below 0,3 rj value were considered to be items in need of correction, and other 41 items were considered to be very good items. Making minor corrections in the items to be corrected, test was applied as it stands.

Prepared aiming at achievements considering students' level of understanding what they read, such as entitling, explaining the reason to choose that title and painting the image in accordance with the colors in the text, KR-20 reliability co-efficient of the test was calculated as 0,89. To provide the validity of the test, opinions of two education experts and three teachers were received.

For the qualitative part of the implementation, three questions were prepared to be asked 10 students chosen by the researcher. Interviews were continued based on the answers to those questions raised during the focus group interviews. Video recordings of the interviews were examined.

1. While playing which game did you have fun in math class? Doing what amused you in this game?

2. What is the difference of playing games in math class from the games you play at home or outside?
3. Would you like to learn in math class by playing or writing? Why?

In the qualitative aspect of the research, interviews were planned by preparing a form of interview including questions to examine students' opinions about the class. Moreover, using an observation form consisting of teacher's observations on the process and examining video recordings of students during the play was planned as well.

Data Analysis

Quantitative data of the research was analyzed by using SPSS 22 packet program. For analyzing quantitative data, descriptive analysis method was preferred. Whether groups are equal or not was determined through applying t-test to the pre-test results of the groups; groups were balanced in terms of age, gender, level of reading comprehension and results of preliminary test. It was seen that pre-test scores of control and experimental groups were equal by the aspect of variables mentioned. Following an 8-week application, experimental and control groups were applied final test evaluating achievements at levels of information, comprehension and implication; determining the difference between the scores of preliminary and final tests of the groups were analyzed via t test. The difference between the averages of final test and pre-test of groups, in other words the difference between their gain score averages, was analyzed via t test. Scores of pre-test information, comprehension and implication levels of groups were covered; analysis of covariance was applied between their continuance score averages. Whether there was a difference or not based on the results of the test, if yes, to what extent there was a difference was tried to be calculated statistically.

In the quantitative aspect of the study, examining video recordings of students' responses to the questions raised in the focus group interviews, examining video recordings and observations of the researcher in the class during the experimental implementation, and analyzing frequency and percentages of students' answers to the open-ended surveys, descriptive analysis method was used. Descriptive analysis is the lowest level and the simplest way of analysis. It is an examination where data is indicated, described, illustrated, explained and reported as it is. In other words, there is no detailed or theory-based decomposition (Sönmez & Alacapınar, 2014). According to this approach, data is interpreted by being summarized based on the themes designated before

Tab. 7: Data Gathered to Respond to Hypotheses and Sub-problems and Analysis Methods Used

Hypotheses	Data Gathered	Analysis Made
1	Pre-test and post- test scores of the experimental group	paired sample t-test
2	Pre-test and post- test scores of the control group	paired sample t-test
3	Achievement (post-test–pre-test) scores of the experimental and control group	independent sample t test
4	Retention test scores of the experimental and control group	Covariance analysis
Sub-Quantitative Problems	**Data gathered**	**Analysis made**
1	Interview data	Descriptive analysis
1	Observation data	Descriptive analysis
1	Video recordings	Descriptive analysis
1	Survey data	Descriptive analysis

(Yıldırım & Şimşek, 2005: 224). Data gathered and analyses carried out to respond to hypothesis and sub-problems of the research are shown in a brief table below (Table 7).

Intervention in the Research

In this study, various plays were covered arranged within the framework of "Numbers" unit for first grade math class. Students' age along with achievements were taken into consideration while preferring plays; four different kinds of games developed by the researcher were used, namely manipulative plays (cube and block games), card games, garden games and competitions. Along with the plays aimed at mathematical skills such as pencil (writing-drawing) activities, affiliating with objects and numbers, counting, addition, subtracting, matching, finding the odd one out, plays involving high level skills related to those skills such as following the instructions, communication, estimation, cooperation, sharing, competing and questioning were preferred. Games were played once or twice a week at the school garden or in the class for about a course hour, and this process continued for 8 weeks. Every week, the game of the former week was repeated or after carrying out evaluation activities related to former games, the game of the current week was played. And in the classes without games

Tab. 8: Applied Games

	Name of the Play	Type of the Play	Objectives
1	Truck-Loading	Manipulative games	Being able to solve basic problems related to addition with natural numbers
2	Number Cubes (Find The Sum)	Manipulative games	Converting certain data related to addition with natural numbers into the desired wording
3	Number Cubes (the sum is the same)	Manipulative games	Being able to solve basic problems related to addition with natural numbers
4	Card Games (find the number pairs)	Card games	
5	Card Games (find the sum)	Card games	
6	Numbers At The Garden (addition)	Garden (outdoor) games	
7	Numbers At The Garden (subtracting)	Garden (outdoor) games	Being able to solve basic problems related to subtracting with natural numbers
8	Darts (mental calculation, addition and subtracting)	Contests (achievement games)	Being able to solve basic problems related to addition and subtracting with natural numbers

played, activities were continued based on the current process in the framework of achievements. At the 9th week, all the games were played once more, so students were reminded of games and subjects. Focus group interviews with 10 students chosen among all students, their opinions about games were asked, and interviews were recorded. The question "While playing which game did you have fun, why?" raised in an open-ended way to the students participating games in the experimental was asked in written form and answers were reported. Then, the final test was applied and efficiency of games was increased. Three weeks later following the appliances of the final test, students were applied the continuance test.

1. Truck-Loading Game

Students were made to play snap cubes beginning from the second semester. In these games, in the framework achievements to make students recognize cubes, counting activities, activity of indicating multitude via a model or activity of explaining multitude of which model was indicated were carried out.

Intervention in the Research 85

Tab. 9: Truck-Loading Manipulative Games Objective and Behaviors

Name of the Play:	Truck-Loading
Type of Play:	**Manipulative Games**

Objective: Being able to solve basic problems related to addition with natural numbers.
Behaviors:
1. In an operation where two natural numbers of which sum does not exceed 20 are added, when sum or one of the addends is given, finding the unknown addend.
2. Determining number pairs of which sum is 10 or 20.

Fig. 5: Truck-Loading Game

In this game, students are separated into groups of pains and told that they are supposed to undertake the duty of loading at a courier company, hence, they are required to calculate the lacking cargo load, and load the truck completely. They are asked to examine the sheets closely, and trying and placing cubes, to observe that the truck is designed to carry 10 cubes vertically.

After beginning the game, studying the operations in the calculation side of the page, they are asked to find the cubes (cargo load) they use. Concrete, semi-concrete and abstract elements are given together in this game. Trying different combinations, students place, count and write numbers, cubes of which sum is 10 in the relevant section, and loading three different trucks, they finish the game. Students are told that as more and complete cargos they have in this game, more points they will get. Students who finish loading all cargos are congratulated, promoted and allowed to visit between-group works (cargos).

2. Number Cubes (Find the Sum)

Students are separated into groups of two (in pairs) and given cubes and worksheets. They are asked to draw a card before beginning. There are some

Fig. 6: Truck-Loading Game

Tab. 10: Number Cubes Game Objective and Behaviors

Name of the Play:	Number Cubes (Find the Sum)
Type of Play:	**Manipulative Games**

Objective:
Being able to turn particular data about addition with natural numbers into desired wording/phraseology.
Behaviors:
1. Indicating two natural numbers with a sum up to 20 via a model.
2. Writing mathematical statement of two natural numbers with a sum up to 20.

addition operations on the card taken. In the operation sheets taken in the first group, the sum of each operation does not exceed 10. This draw has the characteristics of preparation to make students understand the game and the objective. Using cubes in their hands, first they are asked to indicate the operation via a model by getting help from decimal table, next show the model via a symbol with number bubbles, and then write it down as a mathematical statement.

Children are told that cards to be drawn right after these ones will have harder operations, only those who completely finish them will be able to go further to the next steps. Meanwhile, walking around the class, teacher control whether cards and cubes are used in accordance with the rules, if students have mistaken, teacher gives them hints, and wants them to correct their mistakes. Teacher does not correct them himself/herself, does not erase what is

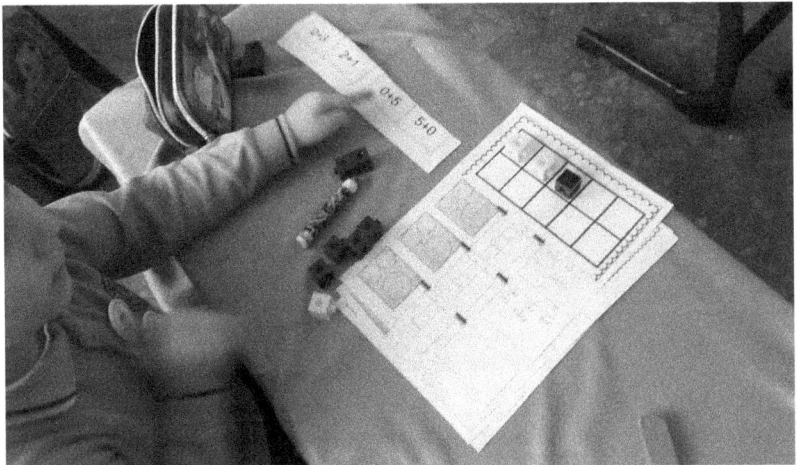

Fig. 7: Number Cubes (Find the Sum)

written and does not relocate cubes. If a student makes huge mistakes, he/she is asked to visit other groups doing it right and come back after watching them. Having heterogeneous groups may be indicated as one of the precautions taken against such problems above. Students with different levels of success should be grouped together.

For the second step, cards with numbers of which sum is up to 20 are drawn by students. Meanwhile, to motivate the class for the game, groups passing to the second step are announced loudly. Again, by the help of cubes, students are asked to fulfill completely the sections of demonstration with a model, number bubbles, demonstration with symbols and writing mathematical statements. And in this step, students are told that cards at the final step are full of surprises, even crayons are involved. Groups finished are announced to the class. They are photographed if possible.

Students arriving at the third step had vaguely learnt the concept of "ten/decimal" in activities and in ten-decimal model. In this step, they are asked to show the numbers of which sum is up to 20, using ten and ones digits. Again, cards are drawn, but students' motivation is tried to be increased saying those cards are the most challenging ones. In this step, they are asked to carry out the operation on the card now without using ten/decimal model, only by using ten and ones digits by the help of cubes.

88 Method

Fig. 8: Number Cubes (Find the Sum)

Involving crayons in this step, concrete cubes are tried to be turned into a semi-concrete form by the help of drawing and painting, and then into abstract forms through symbols and mathematical statement. Thus, while playing with cubes, engaging in painting and writing the operation with symbols, it is aimed that child mathematicize all concrete, semi-concrete and abstract demonstrations representing the operation.

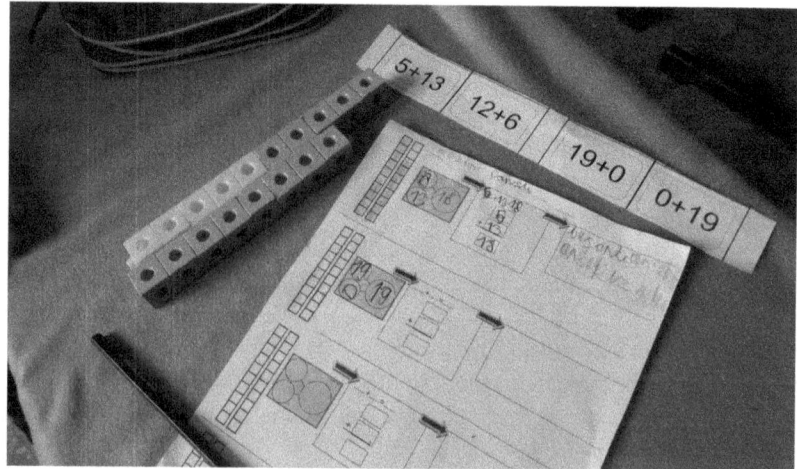

Fig. 9: Number Cubes (Find the Sum)

3. Number Cubes (The Sum Is the Same)

Since students got used to number cubes and playing instructions sheet, now instead of explaining the game, they are told they can understand by reading themselves. So, they will be able to discover the game in the other step as well. In this game with three steps, children are given cubes and worksheets. It is told that instructions will not be given anymore, the rules of the play are written.

Despite of reading the rules, if there are any students who do not start yet, teacher gets closer and tells them to read the instructions loudly. By the help of group pair, the activity starts. For the first model, an operation of which sum does not exceed five is preferred. To make all students recognize the achievement, namely the play, the group started the activity is asked to explain what they have done or understood, so other groups are enabled to continue the play.

Again, cubes, model painting and indicating with operation are included in the play. The greatest motivation to enable children to define those activities as plays is to follow, to obey the rules of the play, namely the completeness of the activity, understanding the activity, students' learning that they are required to perceive the instructions and give feedbacks in a short while because another group may call the teacher any moment and ask for the new instruction sheet.

Another motivation is the free cube play following the final activity. In other words, every play is arranged to be the empowering play for the following one. After finishing the game with three different addition operations, the first groups fulfilled as a whole are congratulated respectively. Each student is given an evaluation sheet and asked to demonstrate what they learn in the game by painting them. In the evaluation activity, students are allowed to have an eye on each other. The purpose here is not evaluation but make them display behaviors to reinforce the achievement. Moreover, they are allowed to sit at different seats/spots in the class and use different crayons.

Tab. 11: Number Cubes Game Objective and Behaviors

Name of the Play:	Number Cubes (the Sum Is the Same)
Type of Play:	**Manipulative Games**

Objective: Being able to solve basic problems related to addition with natural numbers.
Behaviors: Writing and showing the sum is not changed when addends are replaced in an addition operation.

Method

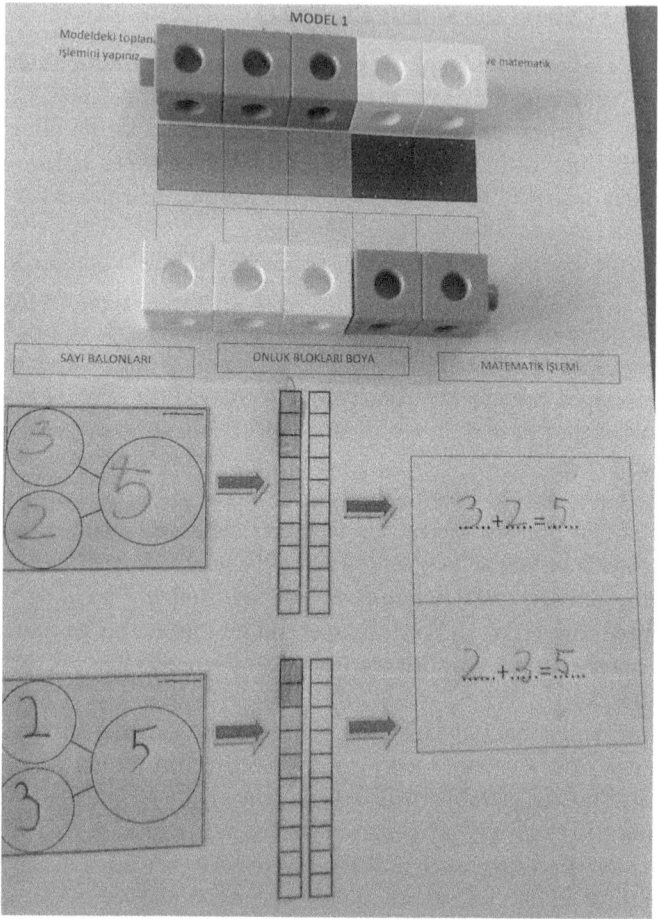

Fig. 10: Number Cubes (The Sum Is the Same)

4. Card Games (Find the Number Pairs)

Students have gained the habit of playing with blocks and cubes. From now on, they want to spend more times with cubes. Of course, it is not because they want to play a math game, but because of considering cubes as Legos (toy blocks), they want to try building many things possible with cubes such as ship, plane, gun, and house, kid, park and so on generally. Even though they use math in

Intervention in the Research 91

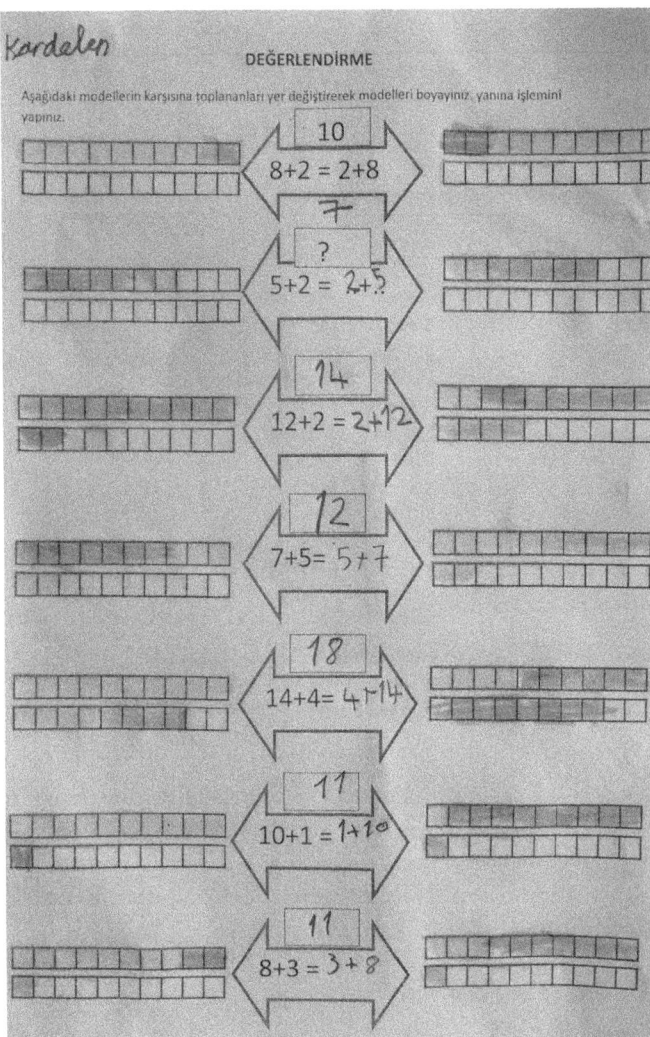

Fig. 11: Number Cubes (The Sum Is the Same)

92　　　　　　　　　　　　　Method

Tab. 12: Card Game Objective and Behaviors

Name of the Play:	(Find the Number Pairs)
Type of Play:	**Card Games**
Objective: Being able to solve basic problems related to addition with natural numbers. Behaviors: Defining number pairs of which sum is 10 or 20.	

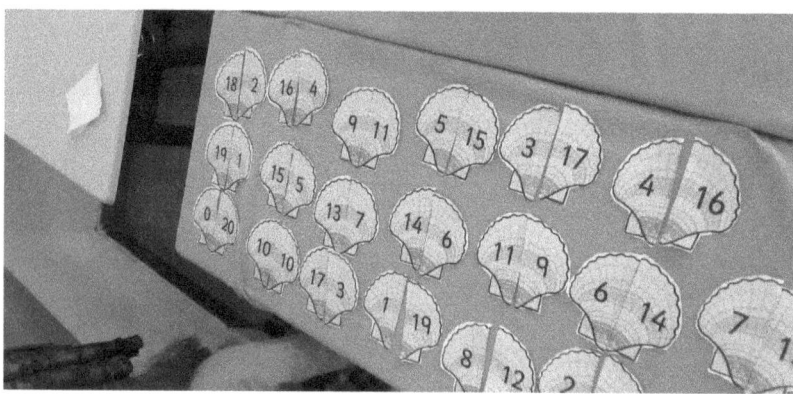

Fig. 12: Card Game (Find the Number Pairs)

their plays, it is the case as it is under the control of teacher, more like using cubes in mathematicizing through magnitude, length and spatial expressions.

In this game, students are separated into groups of four peers, and using the cards with train and oyster shells, they are asked math the numbers of which sum is 10 and 20 out of the cards given disorderly.

5. Card Game (Find the Sum)

Students are separated into groups of four peers and given 40 sheets of which sum is 20. They are told that groups which match the sheets swiftly and find the numbers of which sum is 20 will rank in the competition and have the right to precede the next game. During the game, walking around the classroom, teacher continues to control and correct by raising questions.

The group finishing all the sheets is announced in the classroom. Other groups continue to study. And the group finishing the games is directed to find the numbers of which sum is 19. The game is played during the class. The

Tab. 13: Card Game (Find the Sum) Objective and Behaviors

Name of the Play:	Card Game (Find the Sum)
Type of Play:	**Card Games**

Objective: Being able to solve basic problems related to addition with natural numbers.
Behaviors: Writing natural numbers up to 20 as the sum of two natural numbers.

Fig. 13: Card Game (Find the Sum)

group which completes the highest number of operations is announced in the classroom.

6. Numbers at the School Garden (Mental Calculation; Addition)

Students are taken to the garden and attached with numbers on them up to 20. These numbers are prepared in compliance with classroom size and involving small numbers to have more "number examples" as much as possible. Showing children the small number cards with different numbers on them and saying the number out loud, teacher starts the game. Adjoining with different friends and holding each other's hands, students try to reach the number teacher have told by the means of addition.

Meanwhile, teacher examines the groups emerged, and without revealing the answer, make them reach the right solution. The group finishing the

Tab. 14: Numbers at the Garden Objective and Behaviors

Name of the Play:	Numbers at the Garden (Mental Calculation, Addition)
Type of Play:	**Garden (Outdoor) Games**

Objective:
Being able to solve basic problems related to addition with natural numbers.
Behaviors:
1. Writing natural numbers up to 20 as the sum of two natural numbers.
2. Mental addition of two natural numbers of which sum is up to 20.

Fig. 14: Numbers at the Garden (Mental Calculation, Addition)

game line up at the stairs and show others the addition operation. Teacher says that maximizing the numbers of addends will increase points, and asks students to make groups with addends (students here) of triplet, quadruplet and quintuplet.

The remaining numbers, generally low-level students, are guided to create groups by the help of the teacher. Groups lined up at the stairs are applauded and congratulated by the whole class. The game continues when teacher picks different numbers. As the game proceeds, the numbers on children may be changed.

Fig. 15: Numbers at the Garden (Mental Calculation, Addition)

7. Numbers at the Garden (Mental Calculation, Subtracting)

Children are asked to be gathered with their friends and prepare a subtracting operation. They are told that in the operation, each student will stand for minuend, subtrahend and difference.

Before beginning the game, an example about how to play is displayed. As a distance is covered in the game and the tour finishes, teacher says the difference and subtracting game continues through different number examples. Remaining students are guided and the highest number of subtracting operations is tried to be reached.

Tab. 15: Numbers at the Garden Objective and Behaviors

Name of the Play:	Numbers at the Garden (Mental Calculation, Subtracting)
Type of Play:	**Garden Games**

Objective: Being able to solve basic problems related to subtracting with natural numbers.
Behaviors: Finding mentally the difference of two natural numbers up to 20.

8. Darts (Mental Calculation Addition and Subtracting)

Students are separated into groups of four peers. Students in each group throw three times, respectively. The numbers they hit are written under their group score on the board by a student and group score is calculated by mental addition. The group with the highest number is declared to rank first.

Tab. 16: Darts Game Objective and Behaviors

Name of the Play:	Darts (Mental Calculation Addition and Subtracting)
Type of Play:	**Competitions (Achievement Games)**

Objective: Being able to solve basic problems related to addition and subtracting with natural numbers.
Behaviors:
1. Mental addition of two natural numbers with a sum of up to 20.
2. Finding mentally the difference of two natural numbers up to 20.

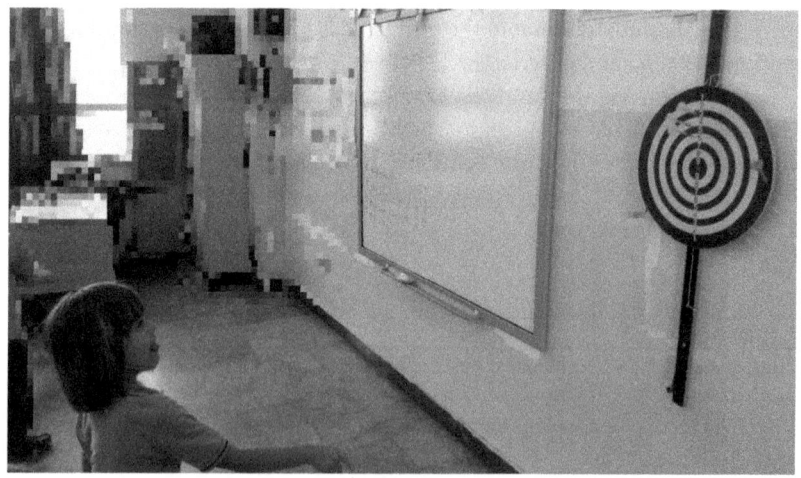

Fig. 16: Darts (Mental Addition and Subtracting)

Fig. 17: Darts (Mental Addition and Subtracting)

Findings

The findings were analyzed in terms of hypotheses of the study in this section whether there was a significant difference in pre-test and post-test scores of experimental and control groups and whether there was a significant difference in achievement, and retention scores between experimental and control groups.

1. Findings Related to First Hypotheses

First Hypotheses of this study is "H1: There is a significant difference between 1st graders' pre-test and post-test scores in experimental group where "the

Tab. 17: Comparison of Experimental Group Regarding Pre-Test–Post-Test Scores

Experimental Group		N	$\bar{x}$	SS	Sd	t	P
Knowledge	Pre-test	27	1,11	,64	26	-4,13	,000*
	Post-test	27	1,67	,48			
Comprehension	Pre-test	27	4,30	2,19	26	-5,79	,000*
	Post-test	27	7,85	3,66			
Application	Pre-test	27	8,89	6,63	26	-10,29	,000*
	Post-test	27	26,37	8,65			
Total	Pre-test	27	14,30	8,44	26	-12,20	,000*
	Post-test	27	36,44	10,42			

* ($p<0.05$)

unit of numbers" is taught through educational plays in math class in primary school?" And to accept or reject this hypothesis before and after the teaching activities were analyzed by paired two sample t-test.

As shown in Table 13, the result was statistically significant (t(54)=-12,20; p= .000<0.05), which revealed a statistically significant difference between pre-test and post-test (total) math achievement test scores of experimental group. Therefore, the first hypothesis was accepted.

2. Findings Related to Second Hypotheses

Second Hypotheses of this study is "H2: There is a significant difference between 1st graders' pre-test and post-test scores in the control group where "the unit of numbers" is **not taught** through educational plays in math class in primary school?" And to accept or reject this hypothesis before and after the teaching activities were analyzed by paired two sample t-test.

As shown in Table 14, the result was statistically significant (t(54)=-6,40; p= .000<0.05), which revealed a statistically significant difference between pre-test and post-test math achievement test scores of control group. Therefore, the second hypothesis was accepted.

3. Findings Related to Third Hypotheses

Third Hypothesis of this study is H3: There is a significant difference between first graders' math achievement in experimental group where "the unit of numbers" is taught through educational plays and the control group where "the unit of numbers" is not taught through educational plays in math class in primary school.

As shown in Table 19, the result was statistically significant (t(54)=-5,26; p= .000<0.05), which revealed a statistically significant difference between

Tab. 18: Comparison Regarding Pre-test–Post-test Scores of Control Group

Control Group		N	$\bar{x}$	SS	Sd	t	P
Knowledge	Pre-test	27	0,93	,67	26	-1,27	,215
	Post-test	27	1,19	,73			
Comprehension	Pre-test	27	3,89	2,11	26	-3,59	,001*
	Post-test	27	5,81	2,94			
Application	Pre-test	27	7,67	6,27	26	-6,12	,000*
	Post-test	27	15,19	7,92			
Total	Pre-test	27	12,48	7,54	26	-6,40	,000*
	Post-test	27	22,19	10,59			

* (p<0.05)

Tab. 19: Comparison Regarding Math Achievement Test Scores of Control and Experimental Groups

Math Achievement		N	$\bar{x}$	SS	Sd	t	P
Knowledge	Control	27	0,26	1,05	52	-1,21	,23
	Experimental	27	0,56	0,69			
Comprehension	Control	27	1,93	2,78	52	-1,99	,51
	Experimental	27	3,56	3,19			
Application	Control	27	7,52	6,37	52	-4,75	,000*
	Experimental	27	17,48	8,82			
Total	Control	27	9,59	8,04	52	-5,26	,000*
	Experimental	27	22,15	9,43			

* ($p<0.05$)

groups in math achievement test scores. Therefore, the third hypothesis was accepted.

4. Findings Related to Fourth Hypotheses

H4: There is a significant difference between first graders' retention in experimental group where "the unit of natural numbers" is taught through educational plays and the control group where "the unit of natural numbers" is not taught through educational plays in math class in primary school?

Table 20 shows that retention mean scores and corrected mean scores are higher in the experimental group in which math lessons was conducted with different types of educational plays.

In Table 21, it was found that the Total retention scores of the students in the experimental and control groups in which *educational plays* was applied in *math class in primary school* were significantly differentiated [F(1, 54)=36,174,

Tab. 20: Adjusted Mean Scores of the Experimental and Control Groups' Retention Scores

Retention	Group	N	Mean $\bar{x}$	Adjusted Mean $\bar{x}$
Retention of Knowledge	Experimental	27	1,74	1,71
	Control	27	0,85	0,87
Retention of Comprehension	Experimental	27	8,04	7,85
	Control	27	6,15	6,33
Retention of Application	Experimental	27	27,19	26,86
	Control	27	13,44	13,77
Total Retention	Experimental	27	36,96	36,27
	Control	27	20,44	21,13

Tab. 21: ANCOVA Results of the Experimental and Control Groups' Total Retention Scores of Corrected for Pre-Test Averages

Dependent Variable	Source	Sum of Squares	df	Mean Square	F	Eta Squared
	Model	5623,510	2	2811,75	33,335 *	,56
Total Retention	Pre-test (reg.)	2572,268	1	44490,74	30,496 *	,37
	Group	3051,243	1	3051,24	36,174 *	,41
	Error	4301,749	51	84,348		
	Total	54416,000	54			

p=0,00 < 0.05].When the pre-test scores were taken under control, it was determined that the retention scores accounted for approximately 41 % of the presence of different teaching practices ($\eta 2$=.41) in the groups that were taught with different teaching practices. As a result, it can be asserted different types of educational plays differed effectively from the teaching methods applied to the control group with respect to retention.

Findings Related to Sub-Problems

Here, findings related to the question "What are students' opinions about teaching through educational plays?" are discussed. In this context, in the semi-structured focus group interview with 10 students in the experimental group following the implementation, students were asked a forehand questions like "What is the difference between playing games in math class and playing at home or outside?" and "How would you like to learn in math class, by playing or by writing? Why?," and their opinions were videotaped. As open-ended questions, students were asked "During which game did you have fun in math class? What amused you in that game?" and they are asked to answer in written forms. Moreover, the processes of playing in class and at garden were videotaped and examined via descriptive analysis method.

When video recordings of students playing in class and at garden were examined, it was clear that students had wanted to spend more times in plays, which provided more movements, enabled them to match with different groups, allowed to manipulate materials in different colors and sizes, resulted in unexpectedly and were under the control of children, and activities in such plays were fast-moving. Students displayed avoidance behavior in plays where the same types of activities with restricted materials were performed and finished the game. They used materials in those plays for different purposes. In plays with responsibilities such as defeating, being defeated, fulfilling the task and cooperative acting, it may be said that children learn mathematical concept by playing for a long time. In planning plays, students' preference of play, their unique world, interests, curiosities and even fears can be considered as important factors in preparing playing environment and materials.

The questions "What is the difference between playing games in math class and at home or outside? How would you like to learn in math class, by playing or writing and why?" were raised in focus group interviews, and students' responses were grouped under various themes. About the difference of math games from daily plays, it was determined that students' opinions centered on the themes "doing math operations," "counting out (designating the "it")" and "toy." While students defined that they designated the it in daily plays of them, and also did the counting-out, generally played with their toy, computers and tablets, they explained their opinions about math games that there were cubes and sheets in math games, darts was used and math games were played in the class and at garden outside.

For the question "How would you like to learn in math class, by playing a game or by writing? Why?" most of the children (6 out of 10) answered as they preferred to play games and other four students stated that they wanted to learn math by writing. Students preferring games stated that writing was tiring and boring, particularly "student M.E." said about this: "I prefer playing because my hands are fatigue while writing." Another student said that he/she could move in the class and spend more time with his/her friends. A student said that they could use different things in games; however, while writing, they only used notebook. When opinions of students who wanted to learn by writing: YA said "I like writing. I cannot solve problem without writing." whereas EC said "My handwriting is very nice, you cannot play all the time." Another student, BU said "I can do math operations easily when I write." In this context, opinions of students in the framework of their approaches to teaching through educational plays in math class can be grouped under the themes of "fun, challenge (effort), communication, competition and success." While answering whether they would prefer different types of plays, the first criteria they stated was that classes must be **fun**. And for another theme, namely "**challenge**," students mostly preferred the plays that they had difficulty in and the ones that required more efforts; it was observed that students participated in plays more during the examinations of class observations. Children, who felt their friends stood by them about **communication** and **cooperation**, focused on the activities more and continued the teacher-student communication constantly. To the contrary of traditional education, the direction of communication in plays was from student to teacher and from student to student. Along with the communicative skill, which enabled to improve high-level skills, students felt free in asking questions and found the opportunity to test together whether their answers were right or not. And when it comes to **competition**, students inclined to behaviors to lift up their individual and group success in plays like darts and numbers at garden. Being motivated more, they continued the activities. It may be said that teacher's taking photos of the completed activities motivated students more. Considering **success theme**, children demanded or waited for another task when they accomplished the tasks. Games or final evaluation activities prepared extra by teacher attracted students with different learning rates to the activities more, and even these kinds of activities were defined as final and difficult aims realized by children, and so games were tried to be completed.

In defining students' opinions about types of math games they played, the question "During which game did you have fun in math class? What amused you in this game?" was raised in written form. Students' answers were shown on a table as positive and negative referring types of plays.

Based on this, while 60 (%85.7) out of 70 different opinions asked from students in written forms are positive, 10 comments are negative. The type of games with the highest number of positive opinions is "**Numbers at Garden**," and then respectively comes truck-loading, darts, cube game, train picking and finding the match. And the ones with the highest number of negative opinions are finding the match and train-picking. In this sense, students' opinions are stated below based on the games they play.

Numbers at Garden:
The Numbers at garden was observed to be the favorite game of students. In the games planned in different weeks for subtracting and adding, students were attached a sheet and asked to do subtracting and adding. Teacher showed the students the card hidden in his/her hand and made the groups that reached the number on the card within the shortest time applauded at the stairs by the whole

Tab. 22: Percentage and Frequency of Students' Opinions about Games Played

Opinion/ play name	Numbers at garden	Truck loading	Cubes game	Darts	Picking trains	Find the match	Total	Percentage %
Positive	17	16	11	10	4	2	60	85,7 %
Negative	2	1	-	1	3	3	10	14,3 %
Total	18	17	11	11	7	5	70	100 %
Percentage	27 %	24 %	16 %	16 %	10 %	7 %	100 %	

group. The fact that students did not know who their partners would be, the number on the card showed by teacher would change and could not be estimated, students focused on the number of themselves and they tried to reach the number shown may improve high-level skills such as mental addition, problem solving, active communication and estimation. Even though students with low level of math success had hard times in the beginning, it was observed that they contributed to the operations and participated in thinking processes as the number changed and successful students encouraged them. Considering the achievements of first grade, students' being able to do mental addition by gathering five different numbers together indicates that game meets high-level skills. Positive opinions of students about this game are as follows:

Femalestudent5: It is a nice game with numbers.
Malestudent2: I liked the number finding game.
Femalestudent7: We played a game at the garden, it was an addition game and enjoyable.
Femalestudent9: ...I liked the cards attached on us because it was fun. For instance, we took photos with whoever could find the number...it was really nice. It was challenging but beautiful.
Femalestudent6:...I would have wanted to play the partner finding game at the school garden, it is fun to try to look for your partner, sometimes your partner might be your best friend. While trying to find your partner, you come closer to each student one by one. You may find wrong partner while trying to find the right one, I liked that so much.
Femalestudent2: Numbers at garden because we play with numbers.
Malestudent5: In addition game, we go to the front of the line, add each other up and find a number, we add that up.
Femalestudent5: Everything was wonderful. I was 14 with Arda, then 8 with Ebrar.
Femalestudent10: We attach number on us in that game at garden. It was so fun while trying to find the numbers on our friends.
Femalestudent4: Adding-up outside was really enjoyable.
Students who expressed negative opinions about the game said materials and sheets used in the game could not attached on them strongly, fell down and paper clips and needles might sting.
Malestudent9: A student was 0, but I wanted to be zero.
Malestudent7: I did not love it because paper clips might hurt you.
Malestudent3: Paper clips and numbers always fell down; I did not love it at all.
This situation brings about the idea that the game can be more flexible and useful when jersey types of numbers or number that can be attached to the outfit are used.

Truck-loading is planned as a game aiming for students to find the numbers of which sum is 10 and 20. In groups of two students, using the cubes, students tried to load the cargo truck completely to fulfill their duties. Students were asked to show the cube activities operationally with symbols on the side section while loading the truck on the sheet with cubes in the play. In the week of play, the computer version of the game projected was played and therefore achievement was reinforced. Students' opinions about games may be stated as follows: game is fun, includes challenges, reinforces subtracting together with addition and develops friendship. What students think about this game is below:

Male student1: I want to play with truck because I was disqualified.
Female student1: It is so fun to play with truck
Female student 2: I like truck-loading game because teacher brings computer and we find it. There are so many nice math games that I want to play all of them.
Male student2: It was very enjoyable to do puzzle. We arranged and located cubes and had fun.
Male student3: Truck-loading is so nice, we place the cubes in the game, I like that.
Male student4: Playing truck-loading is so nice.
Male student5: I like subtracting in truck game.
Male student6: This game is very enjoyable, and there is addition and also what makes me amused is that I am with my friends.
Male student7: I like truck-loading a lot because in this game our friendship is so deep.
Male student8: I do not like it because it is too hard.

Cube game: It is designed as a manipulative game that enables students to write down the solutions of additions, which they drew lots, and the operations representing the models they created using the cubes on proper places in the sheet given. While drawing lots, students had fun a lot and said or showed the operations they picked to each other. It is remarkable that they did this without teacher's references and instructions. Later, students took their seat and modelled the operations with their partner and wrote those down, and they continued with next models. Teachers walking around the class controlled their studies and took photos of the completed studies, therefore supported students. The errors, modelling mistakes or operational mistakes were corrected by teachers' warning "You should check it; you may have something missing there." Motivation that led students to playing was arranged to complete another activity and all other activities. This situation provided the struggle, challenge and participation about which group would be the most successful one in the class. Related to this play, students think that cube games including the warming-up exercises in the preparation stage are fun (ECR), provide communication with friends (EN) and enable to create ten-blocks:

Malestudent1: I liked to play with cubes a lot.
Femalestudent3: Playing with cubes is enjoyable because we build houses.
Malestudent2: I want to play with cube blocks.
Malestudent9: I enjoyed myself most while playing with cubes, I felt happy. Painting was enjoyable.
Femalestudent4: I like to create models a lot. I wish we would do it again.
Femalestudent5: I want to play cubes with my friends.
Malestudent9: I played gladly, it was so fun.

Darts: It is planned as a competing game played individually or in a group, where students throw three darts to the darts placed on the board and try to rise the score of their group by adding up the numbers they hit. About darts, which is chosen by students as their second favorite game, students think as follows:

Femalestudent6: Throwing is fun.
Malestudent6: I liked to play darts.
Malestudent1: I want to play darts, I want adding-up game, but I was defeated in my favorite game.
Malestudent2: We throw the missile, for instance when it hits 40, we add that up and gain points.
Malestudent3: Not knowing about which number we will hit amuses us.
Femalestudent7: I like to play darts because I like it when I win.
Femalestudent8: I liked darts so much.
Femalestudent1: I liked darts, one I hit 12. I had fun since I won.
Malestudent8: I liked it a lot, I had played it before.

The student who expressed a negative opinion stated that he/she missed the shot, could not hit any numbers.

Femalestudent9: I did not like it because I could not win even once.

This situation indicates that it is necessary to have a preliminary for the game or the duration/length of playing should be increased and game should be improved by including more turns. In this kind of competing games requiring certain skills, the fact that students experience the skills in the context of math achievements one by one may increase the quality of and admiration for the game. Meanwhile, groups' competing and uncertainties about which number will be hit may be pointed out as the elements that enable game to be liked and increase the possibility to play it again.

Train picking card game is designed as a game where students match the cards with the numbers of which sum is 10 and 20. In this game that seems to be not liked by students much, it may be said that lack of movement and cards or generally manipulations, being mental, not physical, can decrease the demand for the game. Students' negative opinions are as follows:

Femalestudent5: I did not like the train.
Femalestudent2: I do not like the train game because there is no laughter in it.
Malestudent7: While picking up trains, I would have wanted to finish first with my group.
Femalestudent3: The animal pictures in the train make me scared.
Femalestudent9:...and we really had fun while playing with train cards. And also we had fun while playing with my friends. We played railway modelling. (focus is not numbers)

While students that understand the value of cards and operation can easily complete this game, students are interested in symbols and illustrations on the cards rather than the numbers; this situation brings forward the principle of planning game materials by taking children's interests into consideration. In concern with this situation, this principle is coherent with the principle in resources, which states game and teaching material must be as simple as possible, and must not evoke daily life (Laski et.al, 2015). Even in these games, which does not include competing or defeating elements, it was observed that students generally considered the game as competition and defeating-oriented processes. The most important things to prevent that is set the rules for children in the very beginning and involve in the game together with them. In this way, children can understand that there is no party or the concept of defeating in this game, teacher stands at the equal distance to all students, so they can play more than competing and the mathematical skill underlying the game can be achieved.

Card game **find the match** may be one of the least favorite games among students. Just like in the train game, students tried to match seashells with numbers of which sum is 20. But, maintainability and feedbacks from students indicate that this game did not attract students' attention and must be improved. Students expressed negative opinions because of standing up, restricted objectives and monotype material usage. Based on this, students' opinions are as follows:

Malestudent3: I liked to play with sea shells.
Malestudent5: It is boring since we have to stand up and we cannot play with them outside.
Femalestudent10: I do not like the clam game because it includes counting.
Malestudent6: I do not like card game because there was nothing. There was not number up to 10.

CHAPTER 4 Conclusion, Discussion and Suggestions

Abstract This chapter covers conclusions, discussions and suggestions in relation with the findings of this research, which was conducted to test the effect of teaching mathematics through different play types on first grade students' achievements and retention levels.

4.1. Conclusions Related to the First Hypothesis and Discussions

Conclusions of the first hypothesis of the research, namely "There is a significant difference between pre-test and post-test scores of the experimental group taught through educational plays" are examined. When t test is applied to the total scores of pre-test, post-test, information, comprehension and implementation in the experimental group, a difference with a 0.05 significance level between the scores is observed. Based on these findings, all arrangements carried out during the process and educational games have had a positive impact on students' success at information, comprehension and implementation levels, in other words on learning.

Results are parallel to the research results of Altunay (2004), which used geometry teaching method for the fourth grade, of Yumuşak (2014), which used math teaching method reinforced by games for fourth grades and of Biriktir (2008), which used geometry teaching method via games for the fifth grades in the literature. The results of Yiğit's (2007) research, which used math teaching method supported by computer-aided plays for the second grades, are similar to the results of Dinçer's (2008) research, which used teaching methods through music-aided math games for the second grades. The results of Kılıç's (2010) research, which used math teaching through plays for the first grades, and results of Şirin's (2011) research, which used the method of playing to make preschool students gain the skills of number and operation, support the findings of this study. Results of many researches at secondary school level also support the results of this study. In the results of Songur's (2006) research related to teaching literal expressions and equations through plays and puzzles in math class for the eighth grades, a significant difference between the averages of pre-test and post-test in the experimental group was detected. Both the results of

Gökçen's (2009) research related to teaching common divisions and common multiples through games for the sixth grades and the research of Fırat (2011), which studies the impact of math teaching through computer-aided educational games on conceptual learning for the sixth grade, support the findings of this study, too. It can be said that the results of Canbay's (2012) research related to teaching math through games for seventh grades and Yılmaz's (2014) research for the fifth grades are parallel to each other. It is supported by various research results that teaching math through plays for primary and secondary levels increases the successes of experimental group students.

4.2. Conclusions Related to the Second Hypothesis and Discussions

Results of the second hypothesis of the research, namely "There is a significant difference between pre-test and post-test scores of the control group not taught through educational plays." are examined. When t test was applied to the pre-test and post-test scores of the control group, it was observed that there was a significant difference of 0.05 between the total scores of comprehension and implementation. It was observed that there was **no** significant difference between the scores at information level. During the process in the control group, all arrangements done and the method proposed by the current curriculum had a positive effect on students' success at comprehension and implementation levels, in other words on learning, however, for information level, there was a restricted development. At the end of the process, it can be claimed that students have learnt the achievements in the context of the unit at a certain level. When literature is examined, this result is parallel to the research findings of Yiğit (2007) and Kılıç (2010) for the second grade, Altunay (2004), Biriktir (2008) and Yumuşak (2014) for the fourth grade, where methods proposed by the present program in the control group had a positive impact on students' learning at the elementary level. Similarly, the present method used at secondary school level had a positive impact on the learning of the experimental group students. It may be claimed that the findings of the researches conducted by Songur (2006) for the eighth grade, Fırat (2011) for the sixth grade, Canbay (2012) for the seventh grade and Yılmaz (2014) for the fifth grade correspond to the findings of this study. The fact that methods proposed in the current curriculum increased students' success in the control group has been reinforced by various research results. However, on what level this study is when method of teaching math through plays is taken into consideration will be revealed following the conclusions of the third hypothesis.

4.3. The Conclusions Related to the Third Hypothesis and Discussions

The results of the third hypothesis in the research, namely **"There is a significant difference between the achievement scores of both groups"** are examined. It was seen that there was a significant difference between the total scores of corrected information, comprehension, implementation and gain for the benefit of experimental group. The fact that averages of implementation and total gain scores are higher compared to the control group and there is a significant difference means that teaching method through educational plays used has a positive impact on students' implementation and total gains, in other words on learning. Regarding the level of information and comprehension, there has not been a difference observed in the successes of students in both groups. In the group applied the current method, the averages of information and comprehension gain were positive, however, relatively low. It may be said that the situation here effects success in the control group at a certain level, but is not as influential as teaching method through games. According to this result, it may be claimed that experimental group was more successful than control group at the end of the process. The fact that there are significant differences between groups indicates that teaching through educational games related to daily life including activities such as problem-solving, doing math operations, finding the unknown and manipulation of objects carried by different types of plays is an effective teaching method, and effects the success in the control group to a certain level, but is not as effective as the teaching method through educational games. This conclusion corresponds to the researches of Altunay (2004), Tural (2005), Biriktir (2008), Şirin (2011) and Yumuşak (2014) considering the impact of learning through games in the literature on students' gains. It is more important today for students to perceive the problems they are likely to face in daily life, to solve them and to be in communication. Students who have gained those skills beginning from the early ages will be able to learn without being scared, by loving and playing and even they will be able to teach.

4.4. Conclusions Related to the Fourth Hypothesis and Discussions

When the findings related to the fourth hypothesis of the research, namely the question **"There is a significant difference between the retention test scores of both groups,"** it is seen that there is an important statistical difference between the point averages of implementation, total retention and information corrected based on the pre-test scores of the experimental and control group students. It is understood through the corrected averages that the difference is for the benefit of experimental group. Based on this finding, it can be claimed that experimental process (teaching method through educational plays) substantially affects learning retention in math class for the first grade. While there is not a significant difference concerning the retention averages of comprehension level, it can be said that groups have similar features at this level by the aspect of retention. The result reached for the total retention scores is supported by eta-squared value. In the analysis related to the success variant, eta-squared value calculated within the group indicates that influence quantity has a medium-magnitude, and **being in different groups explains %41 of variability of retention scores**. In the literature, these results of the research are parallel to the researches of Altunay (2004), Songur (2006), Yiğit (2007), Konak (2009), Canbay (2012), Yumuşak (2014) within the context of the impact of teaching math through plays on retention. Learning via plays can be accepted as an influential method in providing retention of what is learnt. Students who do not forget what they learn will be ready for next classes and subjects to be learnt, and therefore they may be more successful. When math skills are taken into consideration, along with the skills such as communicating in daily life, estimating, problem-solving, spatial thinking, arithmetic, skills like being sensitive, making friends, empathy and sharing gain importance. It may be claimed that students who experience and gain those skills in plays and do not forget what they learn for a long while may be successful individuals with high-level skills beginning from the early ages.

4.5. Conclusions Related to Sub-Problem and Discussions

When quantitative data related to the first sub-problem of the research, the question "What are students' opinions on teaching through educational plays?," is examined; it is stated that students have expressed positive opinions, and the games they played more were garden (outdoor) games, competing games, manipulative games and card games, respectively. Duman (2010) points out in his quantitative research that similarly, preschool children generally include social games, and rates of playing games can be designated based on usage of playing outdoor plays, toys, type of play and game behavior characteristics.

According to video-recordings of students videotaped during plays, it is seen that they want to spend more time engaging in plays, which enables to move/act and to be partners with different groups, allows the manipulation of materials in different colors and sizes, has unexpected conclusions, and is under the control of children; activities are not boring and dull, but fast-moving. Badzis (2003) mentions four basic factors that prevent games to be developed and used for teaching in educational implementations. These are contextual obstacles related to game, obstacles related to attitude, structural obstacles and obstacles in implementation. In the video-recordings examined, it is seen that students expressed avoiding behavior in the games with restricted materials and same type of activities and finished the game. In the games where there are responsibilities such as defeating-being defeated, fulfilling the duty and acting in cooperation, it may be said that children played for a long time and learnt mathematical concept. Edo et al. (2009) mentions five different themes based on the class observations and students' products analysis in the context of different symbolic games children play in math class. In this research, it is indicated that students passed through steps like using nonnumeric qualitative expressions during the game (few, a lot and more), using the qualitative meaning of numbers unconsciously (telling the number without thinking on its quality), using qualitative meaning of numbers and nonnumeric expressions (without doing a math operation), did arithmetical operation with numbers (addition and subtracting) and using a calculator.

At the end of the focus group interview with children, their answers to the question What is the difference between playing a game in math class and playing at home or outside? are centered on three themes as "doing a math operation", "counting-out /designating it" and "toys". To the question "How would you like to learn in math class, by playing a game or writing? Why?" majority of children (6 students) answered they preferred playing a game, other four students said they wanted to learn by writing. Students who prefer playing consider writing tiring and boring whereas students who want to learn by writing say game cannot be continuous, and they cannot do math operations.

Students' opinions on teaching through educational plays in math class can be classified under the themes of "fun, challenge (effort), communication, competition and success." In Chen's (2011) research, which studies teachers' point of views and implementations related to learning via plays, it is pointed out that the themes about definitions of play are "Play is fun," "Play is freedom of choice" and "Play is voluntariness." In this research, teachers just like students also mentioned the enjoyable side of this play. Moreover, it can be said that the theme here "Play is voluntariness" supports the themes "challenge" and "communication." Based on inner motivation, in other words voluntary basis, students may interact more, and participate in activities that require more effort. Regarding this, Bergen (1998) mentions that play is a medium tool for communication, and language provides adults a social communication.

Following the development of children's language skills, Bergen also considers play to function as a primary medium, which conveys children's opinions and feelings to others. The theme of competing means that students are not always alone in the competitions they participate. Even though plays are individual or with partners, how other friends playing the same game complete the process gains more importance in games. Regarding that, the control principle of the game Rubin et al. (1983) mentions can be associated to the theme of competing. According to Rubin, in order to consider an activity as a play, inner control must belong to students. While playing a game, especially in a competition, students act according to the flow of the game rather than the directions of teacher and try to accomplish. Regarding the theme of success, just like in the competitions, students want to have feedbacks as they complete and finish the game. Even though math games do not include defeating-being defeated notions, activities that are completed are defined as success and announced to the class, and other play is presented as another accomplishment stage. If games are not success oriented but playing-process oriented, students may have more fun in games and focus on mathematical skills more. Briefly, using plays in math class may have a positive effect on students' success, provide retention and have a positive influence over students' interest in class. Interpreting the findings in the light of relevant literature, some suggestions are made for researches and teachers.

Suggestions

In this study, the impact of teaching through different types of plays on first graders' achievement and retention in math class has been examined. It may be claimed that using this method in math class had a positive influence on students' accomplishment and psychology. In this sense, especially involving first graders in the research, literature review about method and implementation process of this model and post implementation have provided important achievements for the researcher.

1. Using different educational plays in teaching math can be used as an effective method to increase math success of students beginning from preschool or first grade.
2. For math class, enjoyable class activities can be planned in different plays in order to provide continuous learning and enable students to associate daily life to what they learn.
3. This method can be an essential teaching tool to increase students' interests in math and to diminish fear and anxiety of math because when teachers and

students encounter an effective, vivid, dynamic and struggling class, they want to continue that process. Students can also test the ways of fun, diversity of plays and discovering plays in the framework of math achievements. Teachers, researchers, schools, education programmers and educators incur different responsibilities.
4. Teachers should have the chance to start playing activities from day one to change the attitudes toward game.
5. Schools and classes can be equipped with learning and play centers including different materials and playing fields.
6. The numbers of play-oriented educational programs and schools abroad have been increasing. Researchers may attempt to plan a program including play not only in the contest of units but throughout the whole program for elementary school.
7. Educators may improve math games not only physically but also in computer environment.
8. It is observed that studies related to plays in our country are mostly planned via experimental patterns. By carrying out qualitative researches, examples about the implementation may be increased.

This study is carried out with first graders. Next studies may be carried out with different grades and classes. Study may be planned in different disciplines not in context of a single unit but throughout the whole year. Long-termed studies may be conducted increasing the length of study.

Appendix 1: Anticipated Achievements in the Current Curriculum of Math Class for the First Grade

1. Students notice the gathering, adding and increasing meanings of addition.
2. Students display two natural numbers of which sum is up to 20 by the means of a model.
3. Students write the mathematical statement of two natural numbers of which sum is up to 20.
4. Students find addend of two natural numbers of which sum is up to 20.
5. Students show that the sum does not change when addends are replaced in an addition.
6. Students find the addend unknown when one of the addends and sum are given in an operation where two natural numbers of which total sum does not exceed 20.
7. Students designate number couples of which sum is 10 or 20.
8. Students write natural numbers of which sum is up to 20 as the sum of two natural numbers.
9. Students mentally add up two natural numbers of which sum is up to 20.
10. Students establish the problems that require addition with natural numbers of which sum is up to 20.
11. Students solve the problems that require addition with natural numbers of which sum is up to 20.
12. Students notice the factorization, decreasing and deduction meanings of subtracting.
13. Students show the difference of two natural numbers up to 20 via models.
14. Students write mathematical statements related to the difference of two natural numbers up to 20.
15. Students find the difference of two natural numbers up to 20.
16. Students show that when a natural number is subtracted from the same natural number, you get zero.
17. Students solve problems that require subtracting with natural numbers.
18. Students establish problems that require subtracting with natural numbers.
19. Students find the unknown minuend or subtrahend in a subtraction.
20. Students find the difference of two natural numbers up to 20 mentally.

Appendix 2: First Grade Math Class Achievements in Experimental Group Implementation

1. Being able to notice gathering, adding and increasing meanings of addition, choosing and marking them (Information).
2. Being able to display two natural numbers of which sum is up to 20 via a model (Comprehension).
3. Writing the mathematical statement of two natural numbers of which sum is up to 20 (Comprehension).
4. Finding the addend of two natural numbers of which sum is up to 20 (*Application*).
5. Showing that sum does not change when addends are replaced in an addition, choosing and marking them (*Application*).
6. Finding the unknown addend when sum and one of the addends are designated in an addition with two natural numbers of which sum does not exceed 20 (*Application*).
7. Defining number couples of which sums are 10 or 20 (Implementation).
8. Writing natural numbers up to 20 as the sum of two natural numbers (*Application*).
9. Adding two natural numbers up to 20 mentally (*Application*).
10. Assembling problems requiring addition with natural numbers of which sum is up to 20 (*Application*).
11. Solving problems requiring addition with natural numbers of which sum is up to 20 (*Application*).
12. Noticing the factorization, decreasing and deduction meanings of subtracting, choosing and marking (Information).
13. Showing the difference of two natural numbers up to 20 via models, choosing and marking (Comprehension).
14. Writing mathematical statement related to the difference of two natural numbers up to 20 (Comprehension).
15. Finding the difference of two natural numbers up to 20 (*Application*).
16. Showing that you get zero when a natural number is subtracted from the same natural number (*Application*).
17. Solving problems that require subtracting with natural numbers (*Application*).

18. Assembling problems that require subtracting with natural numbers (***Application***).
19. Finding an unknown minuend or subtrahend in a subtraction (***Application***).
20. Finding the difference of two natural numbers up to 20 mentally (***Application***).

Tab. 23: First Grade Mathematics Course Unit Gains and Topics Chart

Gains/Objectives Content	Natural Numbers Learning Area Semantic Knowledge of the Main Concepts Related to Addition Process Sub-Learning Area	Natural Numbers Learning Area; to Be Able to Convert Certain Data Related to the Sub-Learning Area of the Adding Process to the Desired Expression Format	To Be Able to Solve Basic Problems Related to Addition in Natural Numbers	Natural Numbers Learning Area; Semantic Knowledge of the Main Concepts about Subtraction Process Sub-Learning Area	Natural Numbers Learning Area; to Be Able to Convert Certain Data Related to the Sub-Learning Area of the Subtraction Process to the Desired Expression Format	To Be Able to Solve Basic Problems Related to Subtraction in Natural Numbers
	Knowledge	Comprehension	Application	Knowledge	Comprehension	Application
Addition	X (1)					
Converting addition to a model and sentence		X (2)				
Problems requiring addition and its solution			X (8)			
Subtraction				X (1)		
Converting Subtraction to a model and sentence					X (2)	
Problems requiring subtraction and its solution						X (6)

List of Figures

Fig. 1:	The Schema of Play and Learning	40
Fig. 2:	Usage of Play and Learning Varieties in Different Education Levels	43
Fig. 3:	Significant Connections in Understanding Mathematics	51
Fig. 4:	The Lesh Translation Model	52
Fig. 5:	Truck-Loading Game	85
Fig. 6:	Truck-Loading Game	86
Fig. 7:	Number Cubes (Find the Sum)	87
Fig. 8:	Number Cubes (Find the Sum)	88
Fig. 9:	Number Cubes (Find the Sum)	88
Fig. 10:	Number Cubes (The Sum Is the Same)	90
Fig. 11:	Number Cubes (The Sum Is the Same)	91
Fig. 12:	Card Game (Find the Number Pairs)	92
Fig. 13:	Card Game (Find the Sum)	93
Fig. 14:	Numbers at the Garden (Mental Calculation, Addition)	94
Fig. 15:	Numbers at the Garden (Mental Calculation, Addition)	95
Fig. 16:	Darts (Mental Addition and Subtracting)	96
Fig. 17:	Darts (Mental Addition and Subtracting)	97

List of Tables

Tab. 1: Evaluating High- and Low-Loaded Environments 39
Tab. 2: Evaluating Children's Experiences in Terms of Play and Work 43
Tab. 3: Diagram of Quasi-Experimental Design of the Study 80
Tab. 4: Age Distribution of Study Groups 80
Tab. 5: Mathematics Pre-Test Scores of Study Groups 80
Tab. 6: Reading Comprehension Test Scores of Study Groups 80
Tab. 7: Data Gathered to Respond to Hypotheses and Sub-Problems and Analysis Methods Used 83
Tab. 8: Applied Games 84
Tab. 9: Truck-Loading Manipulative Games Objective and Behaviors 85
Tab. 10: Number Cubes Game Objective and Behaviors 86
Tab. 11: Number Cubes Game Objective and Behaviors 89
Tab. 12: Card Game Objective and Behaviors 92
Tab. 13: Card Game (Find the Sum) Objective and Behaviors 93
Tab. 14: Numbers at the Garden Objective and Behaviors 94
Tab. 15: Numbers at the Garden Objective and Behaviors 95
Tab. 16: Darts Game Objective and Behaviors 96
Tab. 17: Comparison of Experimental Group Regarding Pre-Test–Post-Test Scores 97
Tab. 18: Comparison Regarding Pre-test–Post-test Scores of Control Group 98
Tab. 19: Comparison Regarding Math Achievement Test Scores of Control and Experimental Groups 99
Tab. 20: Adjusted Mean Scores of the Experimental and Control Groups' Retention Scores 99
Tab. 21: ANCOVA Results of the Experimental and Control Groups' Total Retention Scores of Corrected for Pre-Test Averages 100
Tab. 22: Percentage and Frequency of Students' Opinions about Games Played 102
Tab. 23: First Grade mathematics Course Unit Gains and Topics Chart ... 121

References

Akman, B. (2002). Okulöncesi dönemde matematik. *Hacettepe Üniversitesi Eğitim Fakültesi Dergisi, 23*(23).

Aksoy, N. C. (2010). Oyun Destekli Matematik Öğretimin ilköğretim 6.Sınıf Öğrencilerin Kesirler Konusundaki Başarı, Başarı Güdüsü, Öz -Yeterlilik ve Tutumlarının Gelişimlerine Etkisi. Yüksek Lisans Tezi, Gazi Üniversitesi, Eğitim Bilimleri Enstitüsü, Ankara.

Akyüz, G. (2013). Öğrencilerin okul dişi etkinliklere ayırdıkları süreler ve matematik başarısı arasındaki ilişkinin incelenmesi. *Elektronik Sosyal Bilimler Dergisi, 12*(46), 112–130.

Alkan, V. (2011). Etkili Matematik Öğretiminin Gerçekleştirilmesindeki Engellerden Biri: Kaygı ve Nedenleri. *Pamukkale Üniversitesi Eğitim Fakültesi Dergisi*, Sayı 29 (Ocak 2011/I), Ss. 89–107.

Altun, M. (2013). Düzenli Eğitsel Oyun Oynayan 11–12 Yaş Grubu Çocuklarda Problem Çözme Becerisinin İncelenmesi. Yüksek Lisans Tezi. Ankara: Gazi Üniversitesi Eğitim Bilimleri Enstitüsü.

Altunay, D. (2004). Oyunla desteklenmiş matematik öğretiminin öğrenci erişisine ve kalıcılığa etkisi. Yüksek lisans tezi. Ankara: Gazi Üniversitesi Eğitim Bilimleri Enstitüsü.

Anthony, G., & Walshaw, M. (2009). Characteristics of effective teaching of mathematics: A view from the West. *Journal of Mathematics Education, 2*(2), 147–164.

Aunola, K., Leskinen, E., Lerkkanen, M. K., & Nurmi, J. E. (2004). Developmental dynamics of math performance from preschool to grade 2. *Journal of Educational Psychology, 96*(4), 699.

Aytekin, H. (2001). Okulöncesi eğitim programları içinde oyunun çocuğun gelişimine etkisi. Yayımlanmamış Yüksek Lisans Tezi, Dumlupınar Üniversitesi Sosyal Bilimler Enstitüsü, Kütahya.

Badzis, M. (2003). Teachers' and parents' understanding of the concept of play in child development and education (Doctoral dissertation, University of Warwick). Retrieved from http://wrap.warwick.ac.uk/2502/.

Baek, Y., Kim, B., Yun, S., & Cheong, D. (2008). Effects of two types of sudoku puzzles on students' logical thinking. In T. Connelly & M. Stansfield (Eds.), *Proceedings of the Second European Conference on Games Based Learning* (pp. 19–24). Academic Publishing Limited UK.

Baroody, A. J. (1989). Manipulatives don't come with guarantees. *Arithmetic Teacher, 37*(2), 4–5.

Bateson, G. (1972). *Steps to an ecology of mind: Collected essays in anthropology, psychiatry, evolution, and epistemology.* San Francisco: Chandler Publishing Co.

Bayazıtoğlu, E. N. (1996). İlkokul 2. Sınıf Hayat Bilgisi Dersinde Eğitsel Oyunlar Erişi ve Kalıcılık Düzeyi. Yayınlanmamış Doktora Tezi. Hacettepe Üniversitesi. Ankara.

Bayraktar, F. & Gün, Z. (2007). Incidence and correlates of internet usage among adolescents in North Cyprus. *Cyber Psychology & Behavior, 10*(2), 191–197.

Bergen, D. (1998). Readings from Play as a Medium for Learning and Development. Association for Childhood Education International, 17904 Georgia Avenue, Suite 215, Olney, MD 20832.

Bergen, D. (2009). Play as the learning medium for future scientists, mathematicians, and engineers. *American Journal of Play, 1*(4), 413–428.

Berlyne, D. E. (1969). Laughter, humor, and play. In Gardner Lindzey & Elliot Aronson (Eds.), *The handbook of social psychology* (Vol. 3, pp. 795–852). Reading, MA: Addison-Wesley Publishing Company.

Biddle, K. A. G., Garcia-Nevarez, A., Henderson, W. J. R., & Valero-Kerrick, A. (2013). *Early childhood education: Becoming a professional.* Thousand Oaks, CA: Sage.

Bilen, M. (1999). *Plandan uygulamaya öğretim.* Ankara: Anı Yayıncılık.

Biriktir, A. (2008). İlköğretim 5. Sınıf Matematik Dersi Geometri Konularının Verilmesinde Oyun Yönteminin Erişiye Etkisi. Yüksek Lisans Tezi, Selçuk Üniversitesi Sosyal Bilimler Enstitüsü, Konya.

Blanton, M. L., & Kaput, J. J. (2005). Helping elementary teachers build mathematical generality into curriculum and instruction. *Zentralblatt für Didaktik der mathematik, 37*(1), 34–42.

Boz, İ. (2014). İlkokul 1. Sınıf Matematik Dersinde Oyunla Öğretim Yönteminin Akademik Başarısına Etkisi. Yüksek lisans tezi. T.C. Zirve Üniversitesi, Sosyal Bilimler Enstitüsi. Gaziantep.

Bragg, L. (2007). Students' conflicting attitudes towards games as a vehicle for learning mathematics: A methodological dilemma. *Mathematics Education Research Journal, 19*(1), 29–44.

Bransford, J. D., Brown, A. L., & Cocking, R. R. (2000). *How people learn* (Vol. 11). Washington, DC: National Academy Press.

Bruner, J. S. (1966). *Toward a theory of instruction.* Cambridge, MA: Belknap Press of Harvard University Press.

Broh, B. A. (2002). "Linking extracurricular programming to academic achievement: Who benefits and why?" *Sociology of Education, 75*, 69–95.

Brush, L. R. (1979). *Why women avoid the study of mathematics: A longitudinal study.* Cambridge, MA: Abt Associates, Inc.

Butler, F. M., Miller, S. P., Crehan, K., Babbitt, B., & Pierce, T. (2003). Fraction instruction for students with mathematics disabilities: Comparing two teaching sequences. *Learning Disabilities Research & Practice, 18*(2), 99–111.

Caldera, Y. M., Culp, A. M., O'Brian, M., Truglio, R. T., Alvarez, M., & Huston, A. (1999). Children's play preferences, construction play with blocks and visual spatial skills: Are they related? *International Journal Behavioral Development, 23*, 855–872. doi: 10.1080/016502599383577.

Canbay, İ. (2012). Matematikte Eğitsel Oyunların 7. Sınıf Öğrencilerinin Özdüzenleyici Öğrenme Stratejileri, Motivasyonel İnançları ve Akademik Başarılarına Etkisinin İncelenmesi. Yüksek Lisans Tezi. T.C. Marmara Üniversitesi Eğitim Bilimleri Enstitüsü.

Carbonneau, K. J., Marley, S. C., & Selig, J. P. (2013). A meta analysis of the efficacy of teaching mathematics with concrete manipulatives. *Journal of Educational Psychology, 105*, 380–400.

Castelli, D. M., Hillman, C. H., Buck, S. M., & Erwin, H. E. (2007). Physical fitness and academic achievement in third- and fifth-grade students. *Journal of Sport and Exercise Psychology, 29*, 239–252.

Catterall, J. S., Chapleau, R., & Iwanaga, J. (1999). Involvement in the arts and human development: General involvement and intensive involvement in music and theater arts. *Champions of Change: The Impact of the Arts on Learning, 1*, 1–18.

Chen, Fong Peng (2011). Children learning through play: Perspectives and Ppactices of early childhood educators in Singapore preschools serving children aged four to six years. University of Leicester. Thesis. https://hdl.handle.net/2381/10119

Cheng, M. F., & Johnson, J. E. (2010). Research on children's play: Analysis of developmental and early education journals from 2005 to 2007. *Early Childhood Education Journal, 37*(4), 249–259.

Christakis, D. A., Zimmerman, F. J., & Garrison, M. M. (2007). Effect of block play on language acquisition and attention in toddlers: A pilot randomized controlled trial. *Archives of Pediatrics & Adolescent Medicine, 161*(10), 967–971.

Clements, D. H. (1999). Subitizing: What is it? Why teach it? *Teaching Children Mathematics, 5*, 400–405.

Clements, D. H., & Sarama, J. (2005). *Math play.* Scholastic Early Childhood Today.

Clements, D. H., & Sarama, J. (2015). Equity and mathematics education. Retrieved from https://www.du.edu/kennedyinstitute/media/documents/equity_2_kennedy_institute_ pages.pdf.

Coşkun, H. (2006). Oyunlarla Dil Öğretimi, Spiele im Sprachunterricht, Learning Languages Through Games, İngilizce, Türkçe, Almanca, CTB Yayınları, Dağıtım Siyasal Kitabevi, Ankara.

David, T., Goouch, K., & Powell, S. (Eds.) (2016). *The Routledge international handbook of philosophies and theories of early childhood education and care.* London: Routledge.

DeGroot, K. (2012). Math Play: Growing and developing mathematics understanding in an emergent play-based environment (Doctoral Dissertation, University of California, San Diego).

Değer, A. Ç. (2012). Çocuk Korolarının Eğitiminde Bir Yaklaşım Olarak Eğitsel Oyun Kullanımının Öğrencilerin Müziksel Erişi Düzeylerine Etkisi. Doktora Tezi Ankara: Gazi Üniversitesi Eğitim Bilimleri Enstitüsü.

Demir, İ., Kılıç, S., & Ünal, H. (2010). Effects of students' and schools' characteristics on mathematics achievement: Findings from PISA 2006. *Procedia-Social and Behavioral Sciences, 2*(2), 3099–3103.

Demir, M. R. (2008). İstasyonlarda Öğrenme Modelinin Hayat Bilgisi Dersindeki Üst Düzey Beceri Erişisine Etkisi. Yayınlanmamış Yüksek Lisans Tezi, Hacettepe Üniversitesi, Sosyal Bilimler Enstitüsü, Ankara.

Demir, M. R., & Yıldızlı, H. (2015). *Middle School Students' Play Preferences and Academic Achievement in Mathematics*, II. Eurasian Educational Research Congress/Ejer Congress, Ankara.

Demirci, N. (2004). İlköğretim I. Kademe Sınıf Öğretmenlerinin Görüşleri Çerçevesinde Oyunla Eğitimin Önemi. Yüksek lisans tezi. T.C. Kafkas Üniversitesi, Sosyal Bilimler Enstitüsü. Kars.

Demirel, Ö. (2004). Planlamadan Değerlendirmeye Öğrenme Sanatı, Ankara: Pegem A Yayıncılık.

Dewey, J. (1933). *How we think.* New York: Heath & Co.

Dewey, J. (1976). *The middle works, 1899-1924* (Vol. 13). SIU press.

Dewey, J. (1990/1915). *The school and society* and *The child and the curriculum.* Chicago: University of Chicago Press.

Dienes, Z. P. (1959). "The Growth of Mathematical Concepts in Children Through Experience." Mathematks III. Offprint from Educ. Research. London: Nat. Found. for Educ. Res. in England and Wales, Newnes Educ. Pub. Co., pp. 9–28.

Dienes, Z. P. (1960). *Building up mathematics.* London, Hutchinson Educational.

Dienes, Z. P. (1963). *An experimental study of mathematics-learning*. Hutchinson of London.

Dienes, Z. P. (1967). Some basic processes involved in mathematics learning. *Research in mathematics education*. Washington, DC: National Council of Teachers of mathematics, 21–34.

Dienes, Z. P. (1971). An example of the passage from the concrete to the manipulation of formal systems. In *The Teaching of Geometry at the Pre-College Level* (pp. 61–76). Springer, Dordrecht.

Dinçer, M. (2008). İlköğretim Okullarında Müziklendirilmiş Matematik Oyunlarıyla Yapılan Öğretimin Akademik Başarı ve Tutuma Etkisi, Yüksek Lisans Tezi, Abant İzzet Baysal Üniversitesi Sosyal Bilimler Enstitüsü, Bolu.

Drew, W. F., Christie, J., Johnson, J. E., Meckley, A. M., Nell, M. L., & Chalufour, I. (2008). A Value-Added Strategy for Meeting Early Learning Standards. *YC Young Children, 63*(4), 38.

Drews, D. (2007). Do resources matter in primary mathematics teaching and learning? In D. Drews & A. Hansen (Eds.) *Using resources to support mathematical thinking*. Exeter, Devon: Learning Matters Ltd.

DuBose, K. D., Mayo, M. S., Gibson, C. A., Green, J. L., Hill, J. O., Jacobsen, D. J., & Donnelly, J. E. (2008). Physical activity across the curriculum (PAAC): Rationale and design. *Contemporary Clinical Trials, 29*(1), 83–93.

Duman, G. (2010). Türkiye Ve Amerika'da Anasınıfına Devam Eden Çocukların Oyun Davranışlarının İncelenmesi "Kültürler Arası Bir Çalışma", Gazi Üniversitesi, Eğitim Bilimleri Enstitüsü, Okul Öncesi Eğitimi Anabilim Dalı. Ankara.

Duman, M. Z. (2008). İnternet Kullanımının Öğrencilerin Sosyal İlişkileri ve Okul Başarıları Üzerindeki Etkisi. *Toplum ve Demokrasi, 2*(3), 93–112.

Duncan, G. J., Dowsett, C. J., Claessens, A., Magnuson, K., Huston, A., Klebanov, P., Paganie L., Feinsteinf, L., Engela, M., BrooksGunng J., Sextonh, H., Duckworthf K., & Japel, C. (2007). School readiness and later achievement. *Developmental Psychology, 43*(6), 1428–1446.

Edo, M., Planas, N., & Badillo, E. (2009). Mathematical learning in a context of play. *European Early Childhood Education Research Journal, 17*(3), 325–341.

Ellis, M. J. (1973). *Why people play*. Englewood Cliffs, NJ: Prentice-Hall.

Ellis, M. J. (1979). The complexity of objects and peers. In B. Sutton-Smith (Ed.), *Play and learning* (pp. 157–174). New York: Gardner Press.

Ercanlı, D. (1997). İlköğretim Okullarının 4.Sınıflarında Dünyamız ve Gökyüzü Ünitesinin Öğretilmesinde Oyun Ve Modellerin Başarıya Etkisi, Yüksek lisans tezi, İstanbul: Marmara Üniversitesi Eğitim Bilimleri Enstitüsü.

Ertürk, S. (1998). *Eğitimde Program Geliştirme*. Ankara: Meteksan Yayınları.

Eveland-Sayers, B. M., Farley, R. S., Fuller, D. K., Morgan, D. W., & Caputo, J. L. (2009). Physical fitness and academic achievement in elementary school children. *Journal of Physical Activity & Health*, 6(1), 99.

Fein, G., & Schwartz, P. M. (1982). Developmental theories in early 41 education. In B. Spodek (Ed.), *Handbook of research in early childhood education* (pp. 82–104). New York: The Free Press.

Fennema, E. (1972). The relative effectiveness of a symbolic and a concrete model in learning a selected mathematics principle. *Journal for Research in Mathematics Education*, 3, 233–238.

Ferholt, B. (2007). Gunilla Lindqvist's theory of play and contemporary play theory. Unpublished paper. Retrieved October, 19, 2011.

Fidan, N. (1996). *Okulda Öğrenme ve Öğretme*. Ankara: Alkım Yayıncılık.

Finlayson, M. (2014). Addressing math anxiety in the classroom. *Improving Schools*, 17(1), 99–115.

Fırat, S. (2011). Bilgisayar destekli eğitsel oyunlarla gerçekleştirilen matematik öğretiminin kavramsal öğrenmeye etkisi. Yüksek Lisans Tezi. Aydıyaman Üniversitesi, Fen Bilimleri Enstitüsü. Adıyaman.

Fisher, K., Hirsh-Pasek, K., Golinkoff, R. M., Singer, D., & Berk, L. E. (2011). Playing around in school: Implications for learning and educational policy. In A. Pellegrini (Ed.), *The Oxford handbook of play* (pp. 341–363). NY: Oxford University Press.

Forman, E. A., & Ansell, E. (2001). The multiple voices of a mathematics classroom community. *Educational Studies in Mathematics*, 46(1–3), 115–142.

Fossa, A. J. (2003). On the ancestry of ZP Dienes's theory of mathematics education. *Revista Brasileira de História da Matemática*, 3(6), 79–81.

Franklin, M. B. (2000). Meanings of play in the developmental interaction tradition. In N. Nager & E. K. Shapiro (Eds.), *Revisiting a progressive pedagogy* (pp. 47–72). Albany, NY: State University of New York.

Frost, J. L. (2010). *A history of children's play and play environments: Toward a contemporary child-saving movement*. New York: Routledge.

Fyfe, E. R., McNeil, N. M., & Borjas, S. (2015). Benefits of "concreteness fading" for children's mathematics understanding. *Learning and Instruction*, 35, 104–120.

Fyfe, E. R., McNeil, N. M., Son, J. Y., & Goldstone, R. L. (2014). Concreteness fading in mathematics and science instruction: A systematic review. *Educational Psychology Review, 26*(1), 9–25. http://dx.doi.org/10.1007/s10648-014-9249-3.

Geertz, C. (1973). Thick description: Towards an interpretive theory of culture. In *The interpretation of cultures: Selected essays* (pp. 3–30). New York: Basic Books.

Geist, E. (2010). The anti-anxiety curriculum: Combating math anxiety in the classroom. *Journal of Instructional Psychology, 37*(1), 24–31.

Gelen, İ., & Özer, B., (2010). Oyunlaştırmanın Beşinci Sınıf Matematik Dersinde Problem Çözme Becerisi Ve Derse Karşı Tutum Üzerindeki Etkisi. *e-Journal of New World Sciences Academy, 5*(1), Article Number: 1C0115.

Gencer, S. L., & Koç, M. (2012). Internet abuse among teenagers and its relations to internet usage patterns and demographics. *Educational Technology & Society, 15*(2), 25–36.

Gilmore, J. B. (1971). Play: A special behavior. In R. E. Herron & B. Sutton-Smith (Eds.), *Child's play* (pp. 311–325). New York: Wiley.

Ginsburg, H. P. (2006). Mathematical play and playful mathematics: A guide for early education. In D. Singer, R. M. Golinkoff & K. Hirsh-Pasek (Eds.), *Play = Learning: How play motivates and enhances children's cognitive and social-emotional growth* (pp. 145–165). New York, NY: Oxford University Press.

Ginsburg, H. P., Lee, J. S., & Boyd, J. S. (2008). Mathematics education for young children: What it is and how to promote it. *Social Policy Report, 22*(1). Society for Research in Child Development.

Ginsburg, H. P., & Seo, K. H. (1999). Mathematics in children's thinking. *Mathematical Thinking and Learning, 1*(2), 113–129.

Gningue, S. (2006). Students working within and between representations: An application of Dienes's variability principles. *For the Learning of Mathematics, 26*(2), 41–47.

Goldstein, J. (2012). *Play in children's development, health and well-being.* Brussels: Toy Industries of Europe.

Goldstone, R. L., & Sakamoto, Y. (2003). The transfer of abstract principles governing complex adaptive systems. *Cognitive Psychology, 46*, 414–466. http://dx.doi.org/10.1016/S0010-0285(02)00519-4.

Gordon, A. M., & Browne, K. W. (2004). *Beginnings and beyond: Foundations in early childhood education* (6th ed.). New York: Delmar.

Gökçen, E. (2009). Ortak Bölenler ve Katlar Konusunun Oyun ile Öğretiminin Başarıya Etkisi.Yüksek Lisans Tezi,On Sekiz Mart Üniversitesi,Sosyal Bilimler Enstitüsü, Eğitim Bilimleri Anabilim Dalı, Eğitim Programları ve Öğretimi Bilim Dalı,Çanakkale.

Gömleksiz, M. (1993). "Kubaşık Öğrenme Yöntemi ile Geleneksel Yöntemin Demokratik Tutumlar ve Erisiye Etkisi", Yayımlanmamış doktora tezi, Adana: Çukurova Üniversitesi.

Gözütok, F. D. (2007). Ögretim _lke ve Yöntemleri. Ankara: Ekinoks Kitabevi.

Grant, S. G., Peterson, P. L., & Shojgreen-Downer, A. (1996). Learning to teach mathematics in the context of systemic reform. *American Educational Research Journal, 33*(2), 509–541.

Gravemeijer, K. (2002). Preamble: From models to modeling. In K. Gravemeijer, R. Lehrer, B. Oers, & L. Verschaffel (Eds.), *Symbolizing, modeling and tool use in mathematics education* (pp. 7–22). Dordrecht: Kluwer.

Groos, K. (1899). *Die Spiele der Menschen*. Рипол Классик.

Gülsoy, T. (2013). 6. Sınıf Öğrencilerinin Kelime Hazinesinin Geliştirilmesinde Eğitsel Oyunların Etkisi. Yüksek Lisans Tezi. T. C. Niğde Üniversitesi Eğitim Bilimleri Enstitüsü.

Güneş, G. (2010). İlköğretim İkinci Kademe Matematik Öğretiminde Oyun ve Etkinliklerin Kullanımına İlişkin Öğretmen Görüşleri. Yüksek Lisans Tezi, Kafkas Üniversitesi Sosyal Bilimleri Enstitüsü Eğitim Bilimleri, Kars.

Gürsakal, S. (2012). Lojistik regresyon analizi ile pisa 2009 öğrenci başari düzeylerini etkileyen faktörlerin değerlendirilmesi. *Süleyman Demirel Üniversitesi İktisadi ve İdari Bilimler Fakültesi Dergisi, 17*(1), 441–452.

Hall, G. Stanley. (1906). *Youth: Its education, regimen, and hygiene*. New York: D. Appleton.

Hanson S. L., & Kraus R. S. (1998). Women, sports, and science: Do female athletes have an advantage? *Sociology of Education 71*, 93–110.

Hart, L. (1992). *Anchor math: The brain-compatible approach to learning*. Village of Oak Creek, AZ: Books for Educators.

Hartmann, W., & Rollett, B. (1994). Play: Positive intervention in the elementary school curriculum. In J. Hellendoorn, R. van der Kooij, & B. Sutton-Smith (Eds.), *Play and intervention* (pp. 195–202). Albany: State University of New York Press.

Haylock, D., & Cockburn, A. D. (2003). *Understanding mathematics in the lower primary years*. London: Paul Chapman.

Hiebert, J., & Wearne, D. (1992). Links between teaching and learning place value with understanding in first grade. *Journal for Research in Mathematics Education, 23*, 98–122.

Hirstein, J. (2008). The impact of Zoltan Dienes on mathematics teaching in the United States. *Mathematics Education and the Legacy of Zoltan Paul Dienes, 2*, 107.

Holt, J. (2009). *How children learn.* Hachette, UK.

Huizinga, J. (1970). *Homo Ludens: A study of the play element in culture.* New York: J & J Harper, 13, 26.

Hughes, B. (2002). *A playworker's taxonomy of play types* (2nd ed.). London: PlayLink.

Hughes, B., & Melville, S. (1996). *A playworker's taxonomy of play tips.* Play Education. Organisation, London: PlayLink.

Humbert, K., & Samuelson, V. (2010). *How First Graders with Low Language Skills Solve Math Word Problems.* Communication Connection: Newsletter of the Wisconsin Speech-Language-Hearing Association.

Hunter, T. & Walsh, G. (2014). From policy to practice?: The reality of play in primary school classes in Northern Ireland. *International Journal of Early Years Education, 22*(1), 19–36. doi: 10.1080/09669760.2013.830561.

Hutt, C. (1971). Exploration and play in children. In R. E. Herron & B. Sutton-Smith (Eds.), *Child's play.* New York: Wiley.

Hutt, C. (1979). Exploration and play (#2). In B. Sutton-Smith (Ed.), *Play and learning* (pp. 175–194). New York: Gardner Press.

Imenda, G. M. (2012). The Promotion and Benefits of Play in Foundation Phase Teaching and Learning. A Dissertation submitted to the Faculty of Education in fulfilment of the requirements of the Degree of Master of Education in the Department of Curriculum and Instructional Studies. University of Zululand, KwaDlangezwa.

Izumi-Taylor, S., Samuelsson, I. P., & Rogers, C. S. (2010). Perspectives of play in three nations: A comparative study in Japan, the United States, and Sweden. *Early Childhood Research & Practice, 12*(1), n1.

Izumi Taylor, S, Rogers, C. S., Dodd, A. T., Kandeda, T., Nagasaki, I., Watanabe, Y., & Goshiki, T. (2004). The meaning of play: A cross-cultural study of American and Japanese teachers' perspectives on play. *Journal of Early Childhood Education, 24*(4), 311–321.

Jordan, N. C., Kaplan, D., Locuniak, M. N., & Ramineni, C. (2007). Predicting first-grade math achievement from developmental number sense trajectories. *Learning Disabilities Research & Practice, 22*(1), 36–46.

Kamii, C. (1991). Toward autonomy: The importance of critical thinking and choice making. *School Psychology Review, 20*(3), 382–388.

Kaminski, J. A., Sloutsky, V. M., & Heckler, A. F. (2008). The advantage of abstract examples in learning math. *Science, 320*, 454–455. http://dx.doi.org/10.1126/science.1154659.

Karabacak, N. (1996). *Sosyal Bilgiler Dersi'nde Eğitsel Oyunlar'ın Öğrencilerin Erişi Düzeyine Etkileri*. Ankara: Hacettepe Üniversitesi Sosyal.

Kılıç, A. Z. (2010). "İlköğretim 1. Sınıf Matematik Dersindeki İşlem Becerilerinin Kazandırılmasında Oyunla Öğretimin Başarıya Etkisi", Celal Bayar Üniversitesi, Sosyal Bilimler Enstitüsü, Manisa.

Kılıç, M. (2007). İlköğretim 1. Sınıf Matematik Dersinde Oyunla Öğretimde Kullanılan Ödüllerin Matematik Başarısına Etkisi.Yüksek Lisans Tezi.T.C. Marmara Üniversitesi Eğitim Bilimleri Enstitüsü Eğitim Bilimleri Anabilim Dalı Eğitim Yönetimi Ve Denetimi Bölümü. İstanbul.

Konak, Ö. (2009). İlköğretğm 6. Sınıf Matematik Dersinde İşbirliğine Dayalı Cebir Öğretiminde Bingo Kartı ve Çalışma Kâğıdı ile Grup Değerlendirmesinin Öğrencilerin Akademik Başarılarına ve Öğrenmenin Kalıcılığına Etkisi. Yüksek Lisans Tezi. Yıldız Teknik Üniversitesi, İstanbul.

Küçükahmet, L. (1995). *Öğretim İlke ve Yöntemleri*. Ankara: Gazi Büro Kitabevi.

Laski, E. V., Jor'dan, J. R., Daoust, C., & Murray, A. K. (2015). What Makes mathematics Manipulatives Effective? Lessons from Cognitive Science and Montessori Education. *SAGE Open, 5*(2), 2158244015589588.

Lazarus, M. (1883). *About the attraction of play*. Berlin, Germany: Dumler.

Lehrer, J. S., Petrakos, H. H., & Venkatesh, V. (2014). Grade 1 students' out-of-school play and its relationship to school-based academic, behavior, and creativity outcomes. *Early Education and Development, 25*(3), 295–317.

Lehrer, R., & Schauble, L. (2002). Symbolic communication in mathematics and science: Co-constituting inscription and thought. In E. D. Amsel, & J. P. Byrnes (Eds.), *Language, Literacy, and Cognitive Development* (pp. 167–192). Mahwah, NJ: Lawrence Erlbaum.

Lesh, R., Cramer, K., Doerr, H., Post, T., & Zawojewski, J. (2003). Using a translation model for curriculum development and classroom instruction. In R. Lesh, & H. Doerr, (Eds.), *Beyond constructivism. Models and modeling perspectives on mathematics problem solving, learning, and teaching*. Mahwah, NJ: Lawrence Erlbaum Associates.

Levine, S. C., Ratliff, K. R., Huttenlocher, J., & Cannon, J. (2012). Early puzzle play: A predictor of preschoolers' spatial transformation skill. *Developmental Psychology, 48*(2), 530–542.

Lieberman, J. N. (1965). Playfulness and divergent thinking: An investigation of their relationship at the kindergarten level. *The Journal of Genetic Psychology, 107*(2), 219–224.

Lindqvist, G. (1995). The aesthetics of play: A didactic study of play and culture in preschools. (Doctoral dissertation). Uppsala: Acta Universitatis Upsaliensis.

Lyons, I. M., & Ansari, D. (2015). Numerical order processing in children: From reversing the distance-effect to predicting arithmetic. *Mind, Brain, and Education, 9*(4), 207–221.

McNeil, N. M., & Fyfe, E. R. (2012). "Concreteness fading" promotes transfer of mathematical knowledge. *Learning and Instruction, 22,* 440e448. http://dx.doi.org/10.1016/j.learninstruc.2012.05.001.

McNeil, N., & Jarvin, L. (2007). When theories don't add up: Disentangling he manipulatives debate. *Theory into Practice, 46*(4), 309–316.

Mehrabian, A. (1976). *Public places and private spaces: The psychology of work, play, and living environments.* New York: Basic Books.

Meier, K. (2015). Overcoming math Anxiety: How Does Teaching Math Conceptually Impact Students Learning math? A research paper submitted in conformity with the requirements for the degree of Master of Teaching Department of Curriculum, Teaching and Learning Ontario Institute for Studies in Education of the University of Toronto, Toronto.

Meletiou-Mavrotheris, M., & Mavrotheris, E. (2012, July). Game-enhanced mathematics learning for pre-service primary school teachers. Paper presented at the meeting of ICICTE, Efstathios Mavrotheris Open University of Cyprus, Cyprus.

Mellon, E. (1994). Play theories: A contemporary view. *Early Child Development and Care, 102,* 91–100.

Miles, M. B., & Huberman, A. M. (1994). *Qualitative data analysis: An expanded sourcebook* (2nd ed.). Thousand Oaks, CA: Sage.

Minton, S. (2003). Assessment of high school students' creative thinking skills: A comparison of dance and nondance classes. *Research in Dance Education, 4*(1), 31–49.

Montessori, M. (1965). *Dr. Montessori's own handbook.* New York: Schocken Books.

Montessori, M. (1973). *From childhood to adolescence; including Erdkinder and the function of the university.* New York: Schocken Books.

Morin, J., & Samelson, V. M. (2015). Count on it: Congruent manipulative displays. *Teaching Children Mathematics, 21*(6), 362–370.

Moyer, P. (2001). Are we having fun yet? How teachers use manipulatives to teach mathematics. *Education Studies in mathematics, 47*(2): 175–197.

Moyles, J. R. (1989). *Just playing?: The role and status of play in early childhood education*. Maidenhead: Open University Press.

National Council of Teachers of Mathematics (2002). *Principles and standards for school mathematics*. Reston, VA: Author.

Neumann, E. A. (1971). *The elements of play*. New York: MSS Information Corp.

Nicolopoulou, A. (1993). Play, cognitive development, and the social world: Piaget, Vygotsky, and beyond. *Human Development, 36*(1), 1–23.

OECD, P. (2009). *Technical Report*. Paris: Author.

Oostermeijer, M., Boonen, Anton, J. H., & Jolles, J. (2014). The relation between children's constructive play activities, spatial ability, and mathematical word problem-solving performance: A mediation analysis in sixth-grade students. *Frontiers in Psychology, 5*, 782.

Ören, Ş. ve Avcı, E. D. (2004). Eğitimsel Oyunla Öğretimin Fen Bilgisi Dersi "Güneş Sistemi Ve Gezegenler" Konusunda Akademik Başarı Üzerine Etkisi, Ondokuz Mayıs Üniversitesi Eğitim Fakültesi Dergisi, 18, 67–76.

Özenç, E. G. (2007). İlk Okuma Ve Yazma Öğretiminde Oyunla Öğretim Yöntemine İlişkin Öğretmen Görüşlerinin İncelenmesi. Yüksek lisans tezi, İstanbul: Marmara Üniversitesi Eğitim Bilimleri Enstitüsü.

Parr, A. (1994). Games for playing. *Mathematics in School, 23*(3), 11–13.

Parten, M. B. (1932). Social participation among preschool children. *Journal of Abnormal Psychology, 27*, 243–269.

Patrick, G. T. W. (1916). *The psychology of relaxation*. Houghton, Mifflin.

Pehlivan, H. (1997). Örnek Olay ve Oyun Yoluyla Öğretimin Sosyal Bilgiler Dersinde Öğrenme Düzeyine Etkisi. Ankara: Hacettepe Üniversitesi Sosyal Bilimler Enstitüsü Eğitim Bilimleri Anabilim Dalı Eğitim Programları Öğretim Bilim Dalı (Doktora Tezi).

Pham, S. (2015). Teachers' Perceptions on the Use of Math Manipulatives in Elementary Classrooms (Doctoral dissertation, University of Toronto), Toronto.

Piaget, J. (1952). *The origins of intelligence in children*. New York: International Universities Press.

Piaget, J. (1962). The relation of affectivity to intelligence in the mental development of the child. *Bulletin of the Menninger Clinic, 26*(3), 129.

Ramani, G. B., & Eason, S. H. (2015). It all adds up learning early math through play and games. *Phi Delta Kappan, 96*(8), 27–32.

Raymond, A. M. (1997). Inconsistency between a beginning elementary school teacher's mathematics beliefs and teaching practice. *Journal for Research in Mathematics Education*, 550–576.

Rosberg, A. M. (2003). Work and play: Are they really opposites? *Opinion Papers* (120), ED 473 151.

Rubin, K. H. (1998). Some "good news" and some "not so good news" about dramatic play. In D. Bergen (Ed.), *Play as a learning medium* (pp. 58–62). Philadelphia: Heineman

Rubin, K. N., Fein, G. G., & Vandenberg, B. (1983). Play. In E. M. Hetherington (Ed.) & P. H. Mussen (Series Ed.), *Handbook of child psychology: Vol. 4. Socialization, personality, and social development* (pp. 698–774). New York: Wiley.

Sadık, R. (2006). The comparison of success levels in four operations and problem solving in natural numbers in primary school grade 4 and 5 students who know how to play chess and who do not (Unpublished master's thesis). Abant Izzet Baysal University, Bolu.

Samelson, V. M. (2009). The influence of rewording and gesture scaffolds on the ability of first graders with low language skill to solve arithmetic word problems. (Doctoral dissertation). Retrieved from: http://ir.uiowa.edu/etd/264.

Samuelsson, I. P., & Carlsson, M. A. (2008). The playing learning child: Towards a pedagogy of early childhood. *Scandinavian Journal of Educational Research, 52*(6), 623–641.

Sarama, J., & Clements, D. H. (2009). Building blocks and cognitive building blocks. *American Journal of Play, 1*(3), 313–337.

Saracho, O. (1986). The Development of the Preschool Reading Attitudes Scale. *Child Study Journal, 16*.

Schwartzman H.B. (1978) Anthropological Play. In *Transformations*. Boston, MA: Springer. https://doi.org/10.1007/978-1-4613-3938-0_1

Sedighian, K., & Sedighian, A. S. (1996). Can educational computer games help educators learn about the psychology of learning mathematics in children. In 18th Annual Meeting of the International Group for the Psychology of Mathematics Education (pp. 573–578).

Shipley, D. (2007). *Empowering children. Play based curriculum for lifelong learning*. Scarborough, Canada: Nelson Education.

Smilansky, S., & Shefatya, L. (1990). *Facilitating play: A medium for promoting cognitive, socio-emotional and academic development in young children*. Gaithersburg, MD: Psychosocial and Educational Publications.

Smith, P. K., & Vollstedt, R. (1985). On defining play: An empirical study of the relationship between play and various play criteria. *Child Development, 56*(4), 1042–1050.

Smith, J., & Cage, B. (2000). The effects of chess instruction on the mathematics achievement of Southern, rural, Black secondary students. *Research in the Schools, 7,* 19-26.

Songur, A. (2006). Harfli İfadeler ve Denklemler Konusunun Oyun ve Bulmacalarla Öğrenilmesinin Öğrencilerin Matematik Başarı Düzeylerine Etkisi, Yayınlanmamış Yüksek Lisans Tezi, T.C. Marmara Üniversitesi Eğitim Bilimleri Enstitüsü İlköğretim Anabilim Dalı İstanbul.

Sowell, E. J. (1989). Effects of manipulative materials in mathematics instruction, *Journal for Research in mathematics Education, 20,* 498-505.

Sönmez, M. T. (2012). 6. Sınıf Matematik Derslerinde Web Üzerinden Sunulan Eğitsel Matematik Oyunlarının Öğrenci Başarısına Etkisi. Türkiye Cumhuriyeti Çukurova Üniversitesi, Sosyal Bilimler Entütüsü, İlköğretim Bölümü. Adana.

Sönmez, V. (2007). *Öğretim İlke ve Yöntemleri.* Ankara: Anı Yayıncılık.

Sönmez, V. (2007). *Program Geliştirmede Öğretmen El Kitabı.* Ankara: Anı Yayıncılık.

Sönmez, V., & Alacapınar, F. G. (2014). Örneklendirilmiş bilimsel araştırma yöntemleri. (3. Baskı) Anı Yayıncılık, 2014. Ankara.

Spidell, R. A. (1985). Preschool teachers' interventions in children's play. Unpublished doctoral dissertation, University of Illinois, Urbana-Champaign.

Spodek, B., & Saracho, O. N. (1987). The challenge of educational play. In D. Bergen (Ed.), *Play: As a medium for learning and development* (pp. 9-22). Portsmouth, NH: Heinemann.

Spodek, B., Saracho, O. N., & Davis, M. D. (1987). *Foundations of early childhood education.* Englewood Cliffs, NJ: Prentice Hall.

Sriraman, B., & Lesh, R. (2007). A conversation with Zoltan P. Dienes. *Mathematical Thinking and Learning, 9*(1), 59-75.

Sylvester, B. N. (1989). First-year teacher usage of manipulatives in mathematics instruction: A case study (Doctoral dissertation, University of North Texas).

Şirin, S. (2011). "Anaokuluna devam eden beş yaş çocuklara sayı ve işlem kavramlarını kazandırmada oyun yönteminin etkisi". Yüksek Lisans Tezi, Uludağ Eğitim Bilimler Enstitüsü: Bursa.

Taşkın, N. (2013). Okul öncesi dönemde matematik ile dil arasındaki ilişki üzerine bir inceleme (Yayınlanmamış Doktora Tezi). Hacettepe Üniversitesi Sosyal Bilimler Enstitüsü, Ankara.

Thomas, J. H. K., & Fox, K. R. (2009).The impact of physical activity and fitness on academic achievement and cognitive performance in children. *International Review of Sport and Exercise Psychology, 2*(2), 198-214.

Thompson, P. W. (1992). Notations, conventions, and constraints: Contributions to effective uses of concrete materials in elementary mathematics. *Journal for Research in Mathematics Education, 23*(2), 123–147.

Thompson, P. W., & Lambdin, D. (1994). Concrete materials and teaching for mathematical understanding. *Arithmetic Teacher, 41*, 556–556.

Thompson, P. W., & Thompson, A. G. (1990). *Salient aspects of experience with concrete manipulatives.* Mexico City: International Group for the Psychology of Mathematics Education,

Tracy, D. M. (1987). Toys, spatial ability, and science and mathematics achievement: Are they related? *Sex Roles, 17*(3–4), 115–138.

Tran, T. M. O. A. (2015). Teachers' Beliefs and How Those Beliefs Affect Manipulative Use in the Classroom. A research paper submitted in conformity with the requirements for the Degree of Master of Teaching Department of Curriculum, Teaching and Learning Ontario Institute for Studies in Education of the University of Toronto.

Tural, H. (2005). İlköğretim matematik öğretiminde oyun ve etkinliklerle öğretimin erişi ve tutuma etkisi. Yüksek lisans tezi, İzmir: Dokuz Eylül Üniversitesi Eğitim Bilimleri Enstitüsü.

Tural Sönmez, M. (2012). 6. Sınıf matematik derslerinde web üzerinden sunulan eğitsel matematik oyunlarının öğrenci başarısına etkisi (Yayımlanmamış yüksek lisans tezi). Çukurova Üniversitesi, Sosyal Bilimler Enstitüsü, Adana.

Turanlı, S. (2012). Oyuna Dayalı Müze Etkinliklerinin Öğrenci Erişi Ve Görsel Sanatlar Dersine Karşı Tutumları Üzerine Etkisi Doktora Tezi Gazi Üniversitesi Eğitim Bilimleri Enstitüsü.

Turner. V. (1969). *The Ritual Process.* Chicago: Aldine.

Uğurel, İ. (2003). Orta Öğretimde Oyunlar ve Etkinlikler ile Matematik Öğretimine ilişkin Öğretmen Adayları ve Öğretmenlerin Görüşleri. Yayınlanmamış Yüksek Lisans Tezi, D.E.Ü. Eğitim Bilimleri Enstitüsü.

Usta, H. G. (2014). PISA 2003 ve PISA 2012 matematik okuryazarlığı üzerine uluslararası bir karşılaştırma: Türkiye ve Finlandiya. Yayımlanmamış Doktora Tezi, Ankara Üniversitesi Eğitim Bilimleri Enstitüsü, Ankara.

Uttal, D. H. (2003). On the relation between play and symbolic thought: The case of mathematics manipulatives. In O. Saracho and B. Spodek (Eds.), *Contemporary Perspectives in Early Childhood.* , Greenwich, CT: Information Age Press.

Van de Walle, J. A., Folk, S., Karp, K. S., & Bay-Williams, J. M. (2011). *Elementary and middle school mathematics: Teaching developmentally* (3rd Can. ed.). Toronto: Pearson.

Van Oers, B. (2010). Emergent mathematical thinking in the context of play. *Educational Studies in Mathematics, 74*(1), 23–37.

Van Oers, B., & Duijkers, D. (2013). Teaching in a play-based curriculum: Theory, practice and evidence of developmental education for young children. *Journal of Curriculum Studies, 45*(4), 511–534.

Vygotsky, L. S. (1967). Play and its role in the mental development of the child. *Soviet Psychology, 5*(3), 6–18.

Vygotsky, L. S. (1978). *Mind in society: The development of higher mental processes*. Cambridge, MA: Harvard University Press.

Weisberg, D. S., Hirsh-Pasek, K., & Golinkoff, R. M. (2013). Guided play: Where curricular goals meet a playful pedagogy. *Mind, Brain, and Education, 7*(2), 104–112.

Wolfgang, C. H., Stannard, L. L., & Jones, I. (2001). Block play performance among preschoolers as a predictor of later school achievement in mathematics. *Journal of Research in Childhood Education, 15*(2), 173–180.

Wood, E. (2008). Everyday play activities as therapeutic and pedagogical encounters. *European Journal of Psychotherapy and Counselling, 10*(2), 111–120.

Wood, E., & Attfield, J. (2005). *Play, learning and the early childhood curriculum*. London: Paul Chapman.

Yiğit, A. (2007). İlköğretim 2.Sınıf Seviyesinde Bilgisayar Destekli Eğitici Matematik Oyunlarının Başarıya ve Kalıcılığa Etkisi. Yüksek Lisans Tezi, Çukurova Üniversitesi Sosyal Bilimler Enstitüsü, Adana.

Yıldırım, Ali & Şimşek, Hasan (2005). Sosyal Bilimlerde Nitel Araştırma Yöntemleri (6. Baskı). Ankara: Seçkin Yayıncılık.

Yılmaz, D. (2014). Ortaokul 5. Sınıf Matematik Dersi Geometrik Cisimler Öğretiminde, Matematik Oyunları Kullanımının Öğrenci Başarısı ve Tutumuna Etkisi. Yüksek lisans tezi Ankara: Gazi Üniversitesi Eğitim Bilimleri Enstitüsü.

Yumuşak, E. Y. (2014). Oyun Destekli Matematik Öğretiminin 4. Sınıf Kesirler Konusundaki Erişi Ve Kalıcılığa Etkisi. Yüksek Lisans Tezi. T.C. Gaziosmanpaşa Üniversitesi Eğitim Bilimleri Enstitüsü. Eskişehir.

www.ingramcontent.com/pod-product-compliance
Ingram Content Group UK Ltd.
Pitfield, Milton Keynes, MK11 3LW, UK
UKHW021324180426
11947UKWH00017B/1420